RENEWALS: 69

DATE DUE

D0568863

WITHDRAWN
UTSA LIBRARIES

PROJECT MANAGEMENT for ENGINEERS

PROJECT MANAGEMENT for ENGINEERS

Milton D. Rosenau, Jr.

LIFETIME LEARNING PUBLICATIONS
BELMONT, CALIFORNIA
A Division of Wadsworth, Inc.

London, Singapore, Sidney, Tokyo, Toronto, Mexico City

Production Management: *Bernie Scheier & Associates*
Copy Editor: *Sylvia Stein*
Illustrator: *John Foster*
Compositor: *Thompson Type*

© 1984 by Wadsworth, Inc. All rights reserved. No part of this
book may be reproduced, stored in a retrieval system, or
transcribed, in any form or by any means, electronic,
mechanical, photocopying, recording, or otherwise, without the
prior written permission of the publisher, Lifetime Learning
Publications, Belmont, California 94002, a division of
Wadsworth, Inc.

Printed in the United States of America

1 2 3 4 5 6 7 8 9 10——88 87 86 85 84

Library of Congress Cataloging in Publication Data

Rosenau, Milton D., 1931–
    Project management for engineers.

    Bibliography: p.
    Includes index.
    1. Engineering—Management. I. Title.
TA190.R53   1984   658.4′04   84-7858
ISBN 0-534-03383-0

LIBRARY
The University of Texas
At San Antonio

Contents

Preface

WHO THIS BOOK IS FOR

This book is for you if you have been trained in a professional skill (such as mechanical, electrical, optical, or chemical engineering) and now find yourself thrust into a position of managing an engineering project in which you've had little or no direct experience. Now you must not only demonstrate your professional skill, but must take responsibility for schedule and budget, using physical and human resources over which you may have little or no real control. You may be an engineer (or scientist) asked to take responsibility for an engineering project, a computer programmer assigned lead responsibility on a software engineering project, or a civil engineer obliged to manage a construction project. No matter what your previous skill, if you're now responsible for getting something done by a specified date and with a limited budget, this book will help you grasp and master key practical skills for successful engineering project management.

Although this book is aimed primarily at the newly appointed manager of an engineering project, technical professionals considering such a position will also find much of value. Project managers with limited engineering project management experience will gain a better understanding or learn new techniques to improve their effectiveness. Functional managers who must support engineering project managers will gain important insights.

Most of the discussion and examples are specific to small and medium-sized projects, rather than to very large projects (such as the space shuttle or trans-Alaska pipeline). However, large projects are normally managed as a collection of smaller, integrated projects; so the book has relevance regardless of project size.

THIS BOOK'S APPROACH TO PROJECT MANAGEMENT

No treatment can or should make trivial a subject as complex as project management, but this book is specifically designed to make it as simple as possible. I provide a step-by-step approach that I have found effective in more than twenty-five years of industrial experience with engineering projects and in teaching thousands of seminar students (mostly engineers), working adults of all ages from a wide variety of industries. By taking you through the engineering project management process in twenty-two steps (Chapters 2–23), the book aims to equip you with detailed tools that you can immediately apply to your first (or next) project and that will help you overcome the pitfalls that typically bedevil the project manager.

USEFUL AND UNIQUE FEATURES OF THIS BOOK

There are many books on single, specialized aspects of project management (for instance, PERT and CPM) and several books on project management in specialized applications (for instance, the construction or aerospace industry), but this book is intended specifically for the beginner who manages engineering projects of any description.

Each short chapter is devoted to a single topic and can be absorbed in one to two hours. Thus, the entire book can be mastered in a single month, making it uniquely useful for the working adult.

The book covers the subject matter chronologically, from a project's beginning to its end, and includes an example project that is developed step by step.

The book will be useful for almost any type of engineering project and has many practical examples. In addition, I use some very basic nonengineering illustrations to assure that a few critical issues are easily grasped, regardless of the reader's specific technical specialization.

Typical problems that may be encountered are highlighted in each chapter.

Useful references in each chapter point to other recent publications if you'd like to read more on any topic.

There are three appendices, containing common abbreviations, a glossary of terms, and helpful checklists for engineering project managers.

HOW THIS BOOK IS ORGANIZED

The step-by-step approach divides the management of engineering projects into five general managerial activities (or processes) and emphasizes the importance of satisfying the three constraints of performance specifications, schedule, and budget.

The first managerial activity is defining the project. Under this topic, I introduce the Triple Constraint and discuss starting successful engineering projects and negotiating and writing contracts.

The second activity is planning. Chapters covering why and how to plan, the work breakdown structure, scheduling tools and time estimating, PERT and CPM network diagrams, planning the budget, the impact of limited resources, and contingency and risk are included in my discussion of planning.

Leading is the third managerial activity, and as part of this topic I discuss how to organize an engineering project, organizing the engineering project team and the support team, the role of the project manager, and some practical tips for engineering project managers.

The fourth activity is monitoring. Monitoring tools, reviews of engineering projects, project cost reports, handling changes, and solving the inevitable problems are covered under this subject.

Under completing, the fifth managerial activity, I tell you how to complete an engineering project and accomplish the final wrap-up.

Milton D. Rosenau, Jr.
Certified Management Consultant
Rosenau Consulting Company
Santa Monica, California

Acknowledgments

I wish to express my thanks to thousands of seminar and executive education students, the various university and other sponsors of these seminars, clients, and my many colleagues in industry who provided so many helpful suggestions and illustrative materials.

The editor, Sylvia E. Stein, has done an outstanding job of clarifying my material. She has found better, shorter, and more lucid ways to express much of what I wrote. I am sure that the readers of this book will benefit greatly because of her work. Whatever errors or confusion may still remain are obviously my responsibility.

It is also a pleasure to acknowledge the fine support provided by my publisher, Alexander Kugushev, and his staff. Neither this book nor my other books would have come into being without his encouragement and help.

Finally, my wife Ellen has been supportive of the time taken out of our lives together to permit this book to be written.

PROJECT MANAGEMENT for ENGINEERS

What Is An Engineering Project?

This chapter differentiates project management from other engineering work and from other managerial activity by examining how projects differ from other activities. It describes the sequence of five managerial activities that characterizes project management.

DISTINGUISHING CHARACTERISTICS OF PROJECTS

Origin

Imagine you are a mechanical engineer and your boss has asked you to manage a materials study project, a chemical engineer who has been asked to set up a pilot plant, a civil engineer who has been asked to manage a construction project, or an electrical engineer who has been asked to manage the relocation of a power supply in an electronics chassis. How would you go about it?

Before reading farther, consider this hypothetical situation. How does this activity differ from others in which you might engage? Is it similar to going to your work place each day? In what ways does it differ?

At this point, you probably want more information about the project. For instance, is it small or large? What do you mean by small or large—the quantity of materials to be studied, the deadline, or the budget? How does project magnitude alter your approach?

This hypothetical project illustrates some of the characteristics that distinguish engineering projects from other activities. Projects originate because something not done before must be done.

Projects are one-of-a-kind undertakings that originate when something has to be done.

Although going to work on some mornings may seem to be a major undertaking, it is not usually considered a project. Going to work is an activity that repeats a prior activity, namely, going to work the day before. This aspect of the definition of a project is not clear-cut. If your project had been to build an amplifier circuit, at some point building a second or third or fourth amplifier circuit ceases to be a project and becomes a repetitive activity (for example, electronic assembly). As a general rule, if the amplifiers are different from one another, then building each is a project. If each amplifier is virtually identical, you have a production line for amplifiers and are not engaged in a project per se.

Product

There are hardware and software projects.

There are many ways to characterize projects. "Hardware project" and "software project" are common terms, depending on whether the final result is a tangible product (hardware) or a computer program accompanied by a report or some other form of documentation (software).

Thus, the product or end result of a project is a second characteristic. A project is not an ongoing activity, but rather an undertaking that ends with a specified accomplishment. In the first hypothetical example, this would be the completion of the materials study project.

Marketplace

Projects can be categorized by their source of sponsorship, as shown in Table 1-1. One of the fringe benefits of being a good engineering project manager is that the tools and techniques applied on the job also have utility for personal undertakings. Figure 1-1 is a time-oriented critical path network (discussed in more detail in Chapter 8) that my wife and I used when we moved from one house to another. Because we only had a thirty-day escrow period, scheduling of many activities in the correct sequences was critical, and we benefited greatly from use of this particular tool, even though this was not an engineering project.

TABLE 1-1. Project Types.

Source	Project
Personal or family	Install automatic lawn sprinklers Repair garage door opener Add a new utility room
Organization sponsored—for the organization itself	Set up a computer-aided design facility Put a pilot plant into operation Commercialize a new product
Customer sponsored—done by your organization under contract	Design and build a new refinery Produce amplifiers for signal-processing equipment Perform radiation leakage tests on an initial production lot
Subcontracted—tasks performed by other organizations for your organization	Build a styling model of your new product Provide a custom microcircuit Install a test fixture for a new test facility
Government sponsored—done by a government entity	Manage a space shuttle project Issue a new $1 coin Conduct tests at the National Bureau of Standards

If the customer is a governmental entity, there will probably be very formal procedures. If your company wishes to conduct a study of new materials for its own information, the process may be very informal.

Whether a project is done under contract to another organization or not affects how formal the process is. But even for a project for your own organization, or even your own department, it's smart to pretend the project is being undertaken for another organization. This pretense helps assure that you will introduce enough formality to manage the project successfully. For instance, a project that is undertaken for another organization requires a contract with a clear, detailed statement of what is to be done, by when, and for how much; such a job specification is sometimes lacking for the initiation of internally sponsored projects.

The nature of the competition is also important. The only materials laboratory in a region is in a favorable situation. A materials laboratory that is one of many in some small city is

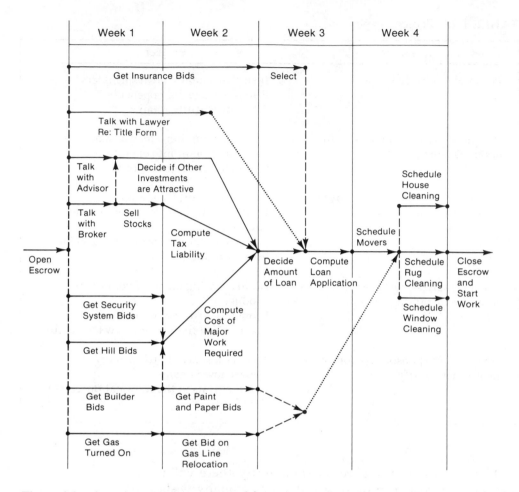

Figure 1-1. A project management tool for a personal project.

in a very different and less favorable competitive position. Inside your own organization (company, government bureau, or university), management can choose to buy the project's deliverable output from a supplier (either standard or custom modified) or ask you to develop it; or management can devote its limited resources to something entirely different, such as building expansion. Thus, you are always in some kind of competition, and it is important to be professional and competent in managing your project.

Once an organization or a project manager has successfully concluded an engineering project, a company-sponsored lens design and fabrication project, for instance, it will be much easier to undertake a second project of that sort. Although there are similarities in managing all kinds of projects, successfully completing a company-sponsored lens design and fabrication project is very different from successfully completing a customer-sponsored telescope project.

In general, projects are done by someone or some organization for someone else or for some other organization.

For whom a project is done affects how it is done.

Project Size

"Program" is sometimes used synonymously with "project." Thus, the expression "program management" may be used interchangeably with "project management." However, programs are now normally considered to be a collection of interrelated projects. Some organizations use "task management" as well. Program management, project management, and task management are generally identical. But programs are usually larger than projects, and projects are usually larger than tasks. Other terms are also used, such as job, work order, and subtask; and these generally refer to still smaller projects. Thus, there is some connotation of size when terms other than project management are used. Nevertheless, the techniques and methodology are essentially the same, differing only in detail. I shall use "project" throughout the rest of this book.

The trans-Alaska pipeline and the space shuttle projects required many years and billions of dollars. The orbiting space telescope has been a long and very costly project, which will gradually shift in the late 1980s to a routine, continuing activity. Conversely, an engineering project to build a breadboard of an amplifier for evaluation may be completed in a few hours or days. Hence, size and complexity do not distinguish a project from another activity. They do, however, affect project success.

Resources

Projects are accomplished by resources, namely, people and things. Many of the required resources are only marginally under the effective control of the project manager. For example, a required lathe may be controlled by a model shop group.

The project manager must organize the correct human resources to take advantage of the available physical resources. Then the project manager has to deal with the constraints and emotional problems inherent in their use while

Managing projects means managing people.

trying to accomplish the project initiator's technical performance goals within the schedule and budget. Managing people is often the most difficult aspect of managing an engineering project, especially for recently appointed managers whose academic training is primarily in engineering. Such people tend to be more comfortable with things and numbers than with people.

Thus, you have to avoid the engineer's propensity to concentrate on the quantitative aspects (for instance, engineering analyses or task budgets)—although these are not unimportant—and instead become more oriented to making things happen through people. Many good engineers make poor project managers because they can't deal with the intangibility of people issues, such as the need to "sell" (and resell) a project to other managers. Some project managers recognize this need but cannot communicate effectively.

The project manager's goal is to utilize organizational resources, which requires internal selling, to achieve project objectives. However, the engineering project manager will often be frustrated by the many other directions the organization seems to be (and often is) moving (see Figure 1-2). These multiple directions arise because of personal aspirations and interests, because of various parochial interests by different components of the organization, and because of many projects being carried out simultaneously. In fact, because you can never be certain which project proposals will be approved (as discussed in Chapter 3), the organization is always somewhat maladapted to the existing work load.

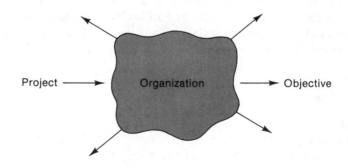

Figure 1-2. The competition for organizational resources impacts a project.

Source: "Project Management." Lecture notes by Milton D. Rosenau, Jr. Copyright © 1981 by the Association for Media-based Continuing Education for Engineers, Inc. (AMCEE). Reprinted by permission.

Because of the tremendous importance of these people issues, I thoroughly discuss organizational options and their influence on resource control and availability in Chapter 12. The rest of Part 3 is devoted to an extensive review of effective actions you can take as project manager to lead both the project and support teams.

The project manager must spend lots of time with people.

THE PROJECT MANAGEMENT PROCESS

Project management requires five different managerial activities and can thus be most simply structured as a five-step process:

1. Defining—defining the project's goals

2. Planning—planning how you and your team will satisfy the Triple Constraint (goal) of performance specification, time schedule, and money budget (The plan depends on the mix of human and physical resources to be used.)

3. Leading—providing managerial guidance to human resources, subordinates, and others (including subcontractors) that will result in their doing effective, timely work

4. Monitoring—measuring or controlling the project work to find out how progress differs from plan in time to initiate corrective action (This often leads to replanning, which may force a goal [definition] change, with a consequent need to change resources.)

5. Completing—making sure that the job that is finally done conforms to the current definition of what was to be done and wrapping up all the loose ends, such as documentation.

The first two steps are not necessarily separate and sequential, except when the project initiator issues a firm, complete, and unambiguous statement of the desired project output, in which case the organization that will carry out the project may start to plan how to achieve it. It is more common to start with a proposed work definition, which is then jointly renegotiated after preliminary planning elucidates some consequences of the initially proposed work definition. In fact, the resources to be dealt with in the leading phase often must be considered before planning can be finished (see

Figure 1-3A). For instance, you might need engineers familiar with carbon fibers if the plan for a materials study project includes the study of that kind of material; whereas you would use a metallurgist if the project were to study only metals.

Similarly, as you will see in later chapters, replanning is almost always required, thus frequently amending the negotiated definition (see Figure 1-3B). Ultimately, the project can be completed when the work that is done satisfies the current requirement (see Figure 1-3C).

Projects require five steps, which may overlap.

Nevertheless, the five-step managerial activity process covers each required action and is a useful conceptual sequence in which to consider all project management, whether engineering or not. Thus, this book is organized according to it.

The Triple Constraint, an extremely important notion for project management, provides the defining parameters of a project. It consists of three dimensions:

1. Performance specification

2. Time schedule

3. Money budget

(In some projects, most commonly research and development or computer software projects, the budget may only be stated in terms of labor hours, rather than dollars. Nevertheless,

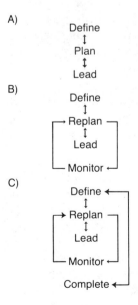

Figure 1-3.   The five activities are interdependent.

labor hours are equivalent to dollars when hourly labor rates are applied.) All projects are defined and characterized by a Triple Constraint.

TYPICAL PROBLEMS

Failure to identify a project for what it truly is usually leads to missed specifications, late completion, and/or a budget overrun. The solution is to recognize that there is a project when something must be done and then to organize to complete the project in the least disruptive way.

The fact that humans are involved in projects and must be worked with is often especially troubling for technically trained project managers. In many cases, this need to work with people is the biggest obstacle that a good technologist encounters. Such technically trained managers expect, but do not get, completely logical or rational behavior from these people (or, for that matter, from themselves). Finally, although many project management tools (for instance, many of the planning and monitoring tools discussed in this book) are completely rational, project management in an overall sense is not an exact science.

Successful project management means meeting the performance specification (that is, objective or technical goals), on schedule, and within the budget.

MATERIALS STUDY PROJECT

I shall use the project introduced here throughout the book to illustrate some of the issues discussed in each chapter. Imagine you are a mechanical engineer, Mel Chase, working for the Successful Project Management Company (SUPROMAC). Your boss, Mike Miller, has asked you to act as project manager on a materials study project a customer (NERESCO or Needs Results Corporation) wishes to have SUPROMAC perform. Because this is a one-time effort, not identical to anything SUPROMAC has done before, project management is clearly going to be applicable.

HIGHLIGHTS   *Projects are temporary undertakings with a specific objective that are accomplished by organized application of appropriate resources.*

Four distinguishing characteristics of projects are origin, product, marketplace, and resources.

Size and complexity do not distinguish projects from other activities.

Project management is the process of achieving project objectives in any organizational framework despite countervailing pressures. This often requires selling and reselling others on the project's importance.

There are five managerial activities in project management: defining, planning, leading, monitoring, and completing.

The Triple Constraint defines a project: performance specification, time schedule, and money budget.

FURTHER READING

R. D. Archibald. *Managing High-Technology Programs and Projects.* New York: Wiley-Interscience, 1976.
   Chapter 2, section 3, is a good short discussion of the distinguishing characteristics of project management.

E. Jenett. "Guidelines for Successful Project Management." *Chemical Engineering* (July 9, 1973), pp. 70–82.
   This is a really fine article with an excellent summary definition of projects. Although it treats the construction industry specifically, this is a valuable short article for all project managers.

Part 1

DEFINING THE GOALS OF AN ENGINEERING PROJECT

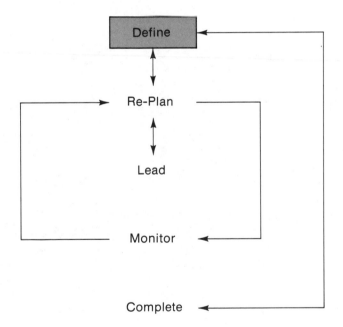

The Triple Constraint

This chapter introduces the concept of the Triple Constraint as a project definition, identifies some of the obstacles to satisfying it, and describes some steps to help achieve it. The consequences of various project outcomes are considered from the point of view of satisfying the Triple Constraint.

THE CONCEPT OF THE TRIPLE CONSTRAINT

Figure 2-1 illustrates the Triple Constraint, a very important concept I shall emphasize throughout the book. Successful project management means accomplishing the performance specifications on or before the time schedule and within the budgeted cost. The cost is usually measured in dollars, francs, marks, or whatever the coin of the realm but may sometimes be measured in the number of labor hours, or perhaps labor hours in each of several labor rate categories such as senior engineer, junior engineer, and so forth. The key point the Triple Constraint illustrates is the need to simultaneously satisfy three independent goals—not just one.

The Triple Constraint defines projects: performance specifications, time schedule, and money (or labor hour) budget.

OBSTACLES TO SATISFYING THE TRIPLE CONSTRAINT

Unfortunately, the Triple Constraint is very difficult to satisfy because most of what occurs during engineering projects con-

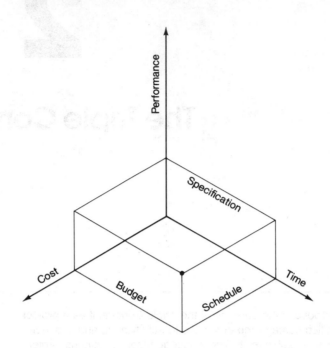

Figure 2-1. The Triple Constraint.

spires to pull the performance below specification and to de-lay the project so it falls behind schedule, which usually makes it exceed budget. The successful project manager is alert to these problems and satisfies the Triple Constraint.

Projects encounter a wide variety of problems. Some of the principal ones are enumerated in the following sections, organized by the dimension of the Triple Constraint most affected. Sometimes only general symptoms are evident, rather than specific problems with a single dimension of the Triple Constraint. Although Figure 2-1 shows these three dimensions to be orthogonal axes, hence mutually exclusive, project management is not that ideal. For instance, the mate-rials study project result may appear poor compared to ex-pectation in relation to the money invested. Similarly, a late project, for example, will usually also overrun the budget. Thus, items listed as causes of time problems may also cause cost problems.

Performance Problems

There are many reasons the performance specification is difficult to achieve. I shall discuss three of them. First, there may have been poor communication between the contractor and the customer. That is, they have different perceptions of the specification or the wording is ambiguous. For instance, "gain" means different things to different people or in different contexts. To an engineer, it may mean amplification, to a thief, booty or spoil. An investor may define gain as increase in value. To a fencer, it may imply advancing the left heel toward the right heel.

A second problem arises because the contractor's or the customer's assumptions have been too optimistic. Their goals may have been too ambitious, which is not uncommon in advanced technology.

Clarify unclear specifications.

Third, the contractor may do a poor design job or make mistakes in executing contract performance. Unfortunately, workers (and managers) make errors occasionally, and these errors may cause a performance deficiency.

Time Problems

Schedule problems arise for several reasons, the most insidious being an overemphasis on the performance dimension at the expense of a balanced view of the Triple Constraint. For instance, engineers who are appointed project managers may tend to concentrate on the technology and to strive for technical innovations or breakthroughs. An electrical engineer may emphasize work on a clever servomechanism design algorithm or use a new programming language rather than expeditiously completing the design with existing capabilities. Such striving is accomplished at the expense of the schedule, and it frequently has unfavorable cost repercussions. To put this another way, "better" is the enemy of "good enough."

Even where a fascination with technology is not overwhelming, technically trained people, such as engineers, tend to assume the performance specification is sacrosanct; whereas they consider it permissible to miss the schedule or budget objectives. Conversely, a customer might be satisfied (if not ecstatic) to achieve 90 percent of the performance specifications, provided both the schedule and the budget are met (for example, on a planetary exploration project, it is vital to be ready by the "launch window," a restricted range of dates, even if one of several nonessential instruments must be omitted).

Technical excellence usually interfaces with meeting the schedule.

A second source of difficulty in meeting the schedule arises because resources are not available when required. These resources may be either equipment (such as lathes or computer hours) or personnel (a well-qualified circuit designer, for example). This absence of planned resources forces the project manager to find substitutes, which may require a subcontract to get some design work done. Or it may mean using marginally qualified people who take longer to complete a circuit design than the well-qualified person previously assumed to be available.

Third, a project can get into schedule difficulty because those assigned to it are not interested in their tasks. In this case, they may choose to work on other things or work half-heartedly on the project.

The time schedule should change if the performance specifications change.

Fourth, schedule problems can arise because the performance specification is raised. For instance, increased efforts that lead to additional work are accepted. A common occurrence that illustrates this is a customer asking for a few extra control switches. The project manager may misperceive this as trivial because a control panel is already being provided. If he or she agrees to provide these extra switches, which were not part of the original proposal, the project manager is agreeing to do additional work (perhaps even modifying the control logic) without changing the schedule (or cost). There is, however, additional work called for to put in the switches, and it does not take many changes of this sort first to produce a one-day schedule slippage and then a one week and so on until the project is in serious schedule difficulty.

Cost Problems

Cost problems arise for many reasons. When an engineering project is in trouble on its time dimension, it will often be in trouble on its cost dimension as well because resources are not being used as efficiently as planned.

A second cause is the "liars contest" that often occurs during contract negotiation (especially if the project is done for another organization). Imagine you have bid $10,000,000 to build a radiometer. During negotiations, you were told that, unless you lower your price to $9,500,000, the contract will be awarded to another company. In your desire to obtain the work, you and your management agree to minor wording changes, which appear to reduce the scope of work a bit and permit you to justify a substantial cost reduction. When you make this kind of price reduction without fundamental work reductions, you have built in a cost overrun at the very beginning of the project. A professional, experienced project manager will never agree to this kind of negotiation unless he or she knows the money will be restored later in contract changes. This is a potential cost problem for which the less experienced project manager must be very alert.

This liars contest also happens inside an organization. In this case, the project has to be sold to upper management, and you are competing with other organization managers for authorization. Chapters 3 and 4 are primarily about doing work under contract for another organization, but they cover much that is important for managers of projects being done inside their own organizations. As I said in Chapter 1, it is smart to pretend your customer and the users are from another organization—and to define the project with all the formality you would use for such an external contract.

A third source of cost difficulties arises because many of the initial cost estimates are simply too optimistic. They do not reflect the inefficiencies that will occur in scheduling resources to perform the work or the fact that less well qualified people may be assigned to do the work.

Occasionally, mistakes are made in the cost estimating. Like design mistakes, these are unfortunate, and careful scrutiny and review can minimize this occurrence.

A fifth reason for cost problems is simply an inadequate cost consciousness on the part of the project management or a failure to have an adequate cost management system. This is never excusable.

Sixth, funding may not occur according to plan, and this may produce problems. Suppose your project is to run for three months and the customer originally proposes to fund you with $100,000 per month (that is, provide progress payments). The first month has gone according to plan and you have received your $100,000. At the start of the second month, your customer informs you they have only $50,000 available for that month but they will have $150,000 available the third month. You will still get $300,000, but you cannot apply the planned resources during the second month and must rush to catch up during the final month. Obviously, this would be a less efficient way to operate and means your costs will be higher than originally planned. When confronted with this sort of refunding proposal by a customer, therefore, the prudent project manager will insist on a budget increase if the performance specifications and schedule are not to be changed.

Funding is required per plan.

Sometimes when you start an engineering project, you and the customer agree that it is impossible to define all the specifications completely. Whenever this happens, it is crucial to agree specifically and in detail on what will be done by a given date. For instance, you may agree to work for one year to establish the complete, final specifications for the project. Then the second phase work can be the rest of the project, which is now aimed at specific, achievable goals. This approach avoids open-ended (thus endless, and hence overrun) projects. The goal you should strive for is a complete, clear, and specific end point.

PROJECT OUTCOMES

A project may end anywhere in the three-dimensional space illustrated by the Triple Constraint. Whether a deviation from the Triple Constraint point is acceptable depends on the project. For example, a project might be late. If the project is to design, test, obtain certification, and then install a vapor recovery system by a legislated date, lateness would be inappropriate. As suggested earlier, schedule compliance is also the overriding concern for planetary launch.

In other situations, the budget may be the most important issue. When a contractor accepts a fixed price contract to deliver a working piece of hardware, he or she will lose money if the budget is overrun. Conversely, on a cost reimbursable contract, the customer must pay for the overrun. In either situation, ability to pay is crucial.

Third, there are situations in which the performance specification may not be missed. Once again, the project specifics will determine whether this is crucial. In the early 1960s, before spacecraft could carry large instruments, the Stratoscope II project was undertaken. The goal was to carry a large telescope by balloon to eighty thousand feet, where it would be above almost all the earth's atmosphere, permitting the optical resolution of fine detail in distant nebulae. To accomplish this goal, the telescope's primary mirror had to be virtually perfect. Because the project could succeed only with such a mirror, there was no point in flying the balloon-borne telescope until this performance specification had been met. This required both more time than the original schedule and more money than the original budget. In other cases, there may be some law that must be complied with.

TYPICAL PROBLEMS

The project initiator's emphasis is almost always unclear initially, and project personnel tend to assume their own biases in ranking the relative importance of each dimension. This can easily lead to a disastrous outcome, which can be avoided by adequate discussions between the customer and the contractor at the project's inception.

The other major problem, mentioned earlier, is the myopic attention to the performance dimension by technical personnel. It can be overcome, or at least reduced, if the engineering project manager clearly conveys the customer's emphasis and its rationale.

Project specifics determine the relative importance of each dimension of the Triple Constraint.

MATERIALS STUDY PROJECT

In discussion within SUPROMAC about the NERESCO materials study project, Mel Chase realizes there are easily half a dozen materials that might be relevant to NERESCO's needs and that SUPROMAC might therefore reasonably examine. He feels this will require about one year and cost $1 million.

Discussions with NERESCO reveal that they require the study results in six months and their budget cannot exceed $500,000. Thus, some of the possibilities for Mel Chase to consider are the following:

Try to convince NERESCO that he is right and more materials should be studied, taking longer and costing more than they want.

Refuse to undertake the project.

Divide the project into several phases. The early phases can be very quick and low cost, requiring joint SUPROMAC-NERESCO review upon conclusion to set goals for later phases.

continued

> Explain the risk to NERESCO in studying only a few materials, and jointly work with them to choose the specific (fewer) materials to be studied within their time and budget constraints.
>
> One of the last two possibilities is probably preferable in this case.

HIGHLIGHTS

The Triple Constraint defines all projects.

The Triple Constraint consists of performance specifications, a time schedule, and a money or labor hour budget.

Obstacles that prevent satisfying the Triple Constraint are not mutually exclusive.

Project specifics determine the relative importance of each dimension of the Triple Constraint.

Adequate and clear discussions among the customer, the project manager, and the technical personnel can help avoid many common problems.

FURTHER READING

R. D. Archibald. *Managing High-Technology Programs and Projects,* New York: Wiley-Interscience, 1976.
   Chapter 1, section 5, has a brief discussion of some causes of poor project outcomes.

V. G. Hajek. *Management of Engineering Projects.* New York: McGraw-Hill, 1977.
   Chapter 1, section 1, has a short discussion of some disciplines and functions needed to be successful with projects.

H. Kerzner. *Project Management: A Systems Approach to Planning, Scheduling and Controlling.* New York: Van Nostrand Reinhold, 1979.
   Chapter 1, section 1, enumerates the social and environmental issues impacting project success.

P. W. Metzger. *Managing a Programming Project,* 2nd ed. Englewood Cliffs, NJ: Prentice-Hall, 1981.

> *Chapter 2, pages 16–49, provides a good discussion of many specific issues to be considered in the definition phase of a software project.*

3

How to Start Successful Engineering Projects

Proposals bridge the defining and planning phases of engineering projects. This chapter examines the strategic issues that govern writing successful proposals. Although this chapter focuses on proposals written for an external organization, all the same principles apply to proposing to undertake projects within your own organization. Project proposals constitute the definition embodied in the contract or work authorization. A good proposal includes a thorough plan for work performance that embodies the Triple Constraint. This chapter also discusses the proposal process in detail.

STRATEGIC ISSUES

Framework

Figure 3-1 illustrates the strategic framework for obtaining winning projects. It does not matter whether the projects originate outside the organization by a customer or within the organization. However, many organizations dissipate their energies in preparing losing proposals, which

1. Cannot win the competition
2. Will have an unfavorable outcome with regard to the Triple Constraint
3. Can be successful but are insignificant or irrelevant to the proposing organization

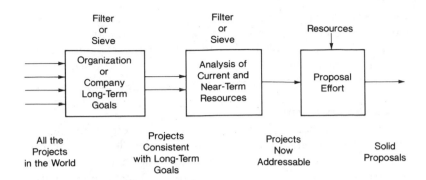

Figure 3-1.   Strategy to obtain winning projects.

Therefore, the basis of a successful strategy is to filter out losing projects. These include projects inconsistent with the organization's long-term goals or with the current and near-term resources within or otherwise available to the organization. Such filters might reject consumer project efforts in an industrial product company. Similarly, the filters could reject a fixed price contract for a technical development in a conservative company that will not normally undertake a fixed price contract for something not previously accomplished.

Filtering out the huge number of possible projects the organization might have addressed leaves a much smaller number of projects that are appropriate for the organization to consider. It can then address some or perhaps all of these in proposal efforts to which it applies adequate and appropriate resources. The result of this process is that an organization submits only very well founded proposals for consideration. This result is most likely to be achieved if the organization has a careful review process, often called the "bid/no bid" decision.

In summary, avoid projects that are inconsistent with your organization's long-term goals or current and near-term resources, unlikely to win the proposal competition, unlikely to satisfy the intended Triple Constraint, or insignificant or irrelevant.

Concentrate on meaningful prospective winners.

Bid/No Bid Decisions

The decision whether to bid on a proposal opportunity, whether to an external organization or within the organization itself, must be taken within the context of the organization's strategic framework. This framework is of course specific to the organization at that particular time. The company that rejects consumer product projects today may have the interest and capability to undertake these projects five years from now. There are many issues involved in this decision, four of which I discuss.

The Requirement

Clear criteria are vital.

The first issue is whether there is a real requirement. If a requirement exists, it is important to decide whether the funds are really available. It is possible for there to be a real requirement but no funds. For instance, many isolated rural villages in undeveloped nations could use an electric power generation system, but these villages and the nations of which they are a part cannot afford to pay for their installation. Contractors occasionally seek subcontractor proposals for "window dressing" to satisfy someone or otherwise justify an alternative approach they propose to take. Clearly, it is not worth responding to such requests for a proposal because the effort will not result in a winning project.

It is also important to examine the priority or importance of the proposed project. In this connection, the relationship of a particular proposal opportunity to present and future programs is also an important issue. Many small "paper study" projects, not attractive per se, actually are very attractive because they can lead in the future to large production programs.

Project Value

The project's social, ecological, and energy impacts might be highly significant to an organization in deciding whether or not to bid on a project. Presumably, a project to build the prototype of a device to clean up oil spills economically would be more attractive than a project that endangers the earth's ozone layer. Similarly, a project that offers the opportunity to apply important new technology or otherwise enhance the organization's reputation might have high nonmonetary value. Conversely, the expected sales of a commercial new product project effort and the profits of such an effort would also be significant issues.

Finally, many projects are merely a hidden obligation for an organization to accept future financial commitments for new capital or facility investments. This must be discovered before any money is spent so as to be certain that future major financial commitments are within the organization's resources and abilities as well as being consistent with the project's prospective value.

Response Ability

The central issue here, also illustrated in Figure 3-1, is the organization's present capability first to prepare a winning proposal and second to perform the proposed work. If some capabilities are not actually present, there must be a viable plan to make them available when they are needed.

The organization must have the ability.

Winning the Competition

First, an organization must ask whether there was advance information about the project available to it. This is particularly true about efforts arising from a customer organization and being presented to a contracting company as a request for a proposal (RFP), but it is also relevant for efforts within an organization. Lack of advance information often indicates that someone else has a head start or the request was hastily created and lacks substance.

A second issue concerns the customer. Is it an individual (yourself, a friend, or someone else) or a commercial organization? If the latter, is it your organization or an external organization? If it is your own company, has your superior ordered you to carry out the project or must your project proposal compete with other project proposals that management is considering for funding? If the customer is a governmental organization—a city, county, state, or federal (domestic or foreign) entity—there will probably be detailed specifications, formal quality standards, perhaps the necessity for surety bonds, and very often rigid and formalized inspection procedures. Who are the key personnel within the customer organization? Are they known to your organization? What history do you have with them or with their organization? Is your organization's reputation with the prospective customer favorable?

Your organization must have enough money to write the proposal and to sustain the postproposal selling and negotiating efforts. Therefore, you must know whether money to invest in this kind of activity is available. You must expect the project to earn more money than the proposal costs because you will not win every job on which you propose.

There are situations in which you will be the sole source recipient of an RFP, for instance, when your boss tells you to carry out a project within your organization. There is thus no competition, but how you perform the job is still important. It may be better to decline an effort when it is offered sole source if you are convinced the performance you can provide will be at best marginal.

All competition has to be analyzed. Some relevant issues are the competition's technical and managerial competence, its ability to produce the requested project output, an estimate of its interest in the particular type of project, its need (or degree of "hunger"), and its prior relationship with the customer or organization.

In summary, avoid projects that are inconsistent with your organization's long-term goals, inconsistent with your organization's current and near-term resources, unlikely to win the proposal competition, unlikely to satisfy the intended Triple Constraint, or insignificant or irrelevant.

THE PROPOSAL PROCESS

Although you will not win every proposal, you must be willing and able to satisfy any commitments you propose to undertake.

The proposal process entails more than writing the proposal (see Figure 3-2). That's the meat of the proposal sandwich. It also requires preparation work and postsubmission work. The goal of the preproposal work is to learn about the customer's problem and bias, which allows you to set a proposal theme. The proposal theme provides a focus to direct everyone contributing to the proposal, which increases the odds of producing a coherent, winning proposal. Thus, the overall proposal process includes the following:

1. Authorization (which formally considers the issues previously discussed)

2. Selection of a dominant theme

3. Preparation of the statement of work

4. Development of a plan to satisfy the Triple Constraint (an effort for which checklists may be helpful)

5. Adjustment to remove inconsistencies and inadequacies.

6. Approval

7. Submission

8. Postsubmission follow-up, including presentations and contract negotiations

Authorization

The proposal process frequently starts before an RFP, in which case the effort is frequently called the preproposal effort. Regardless of when the proposal process is initiated, this activity must be authorized. A form such as that in Figure

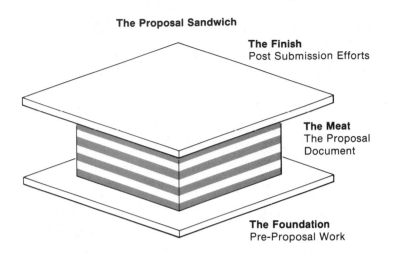

Figure 3-2. The proposal sandwich.

Source: "Project Management." Lecture notes by Milton D. Rosenau, Jr. Copyright © 1981 by the Association for Media-based Continuing Education for Engineers, Inc. (AMCEE). Reprinted by permission.

3-3 can be used for this purpose. In preparing a proposal, an organization is going to commit a certain amount of effort and money to it. This investment should be made only when it seems that the opportunity then available has a good chance of paying off and only if it is consistent with the organization's goals.

Another point to consider at the time of proposal authorization is the individual who will manage it. Ideally, the proposal manager should be the intended project manager. Often this ideal is not achieved. A proposal manager who knows that he or she will not be the project manager is likely to make commitments that are hard (or impossible) to meet, especially when the proposal manager's performance is judged on his or her ability to obtain (as opposed to manage) new business.

A proposal itself can be considered a project with a Triple Constraint, in which case the performance objective is submission of a winning proposal and price, in accordance with the required submission schedule, for a cost acceptable in view of the probable (financial) return to your organization.

Consider, then begin.

Theme Fixation

Except in those very formal RFPs in which contact with the prospective customer is either prohibited or heavily con-

PROPOSAL AUTHORIZATION	NUMBER	REVISION
TITLE		

JOB			
	PERFORMANCE REQUIRED		
	ESTIMATED STARTING DATE	ESTIMATED DURATION	
	ESTIMATED BID PRICE	ESTIMATED SUBCONTRACT TO OTHERS $ %	
	IS JOB FUNDED?	WHAT IS FOLLOW-ON POTENTIAL?	
	ESTIMATED NEED FOR CAPITAL AND FACILITY EXPENSE IF JOB IS OBTAINED		
	CUSTOMER ORGANIZATION		
	KEY CUSTOMER PERSONNEL		
	CONTRACT FORM	SPECIAL CONSIDERATIONS	SECURITY CLASS

COMPETITION	
	COMPETITORS
	COMPETITORS' STRENGTHS
	SIGNIFICANCE TO COMPETITORS IF THEY LOSE
	OTHER COMPETITOR WORK FOR CUSTOMER
	OUR UNIQUE ADVANTAGES

PROPOSAL		
	WHAT IS TO BE SUBMITTED?	
	DUE DATE	PROPOSAL COST (DETAIL BELOW)
	PROPOSAL MANAGER	OTHER KEY PROPOSAL PERSONNEL

EFFORT / ACTIVITY	PROJECT DEPT. (HOURS)	SUPPORT GROUP A (HOURS)	SUPPORT GROUP B (HOURS)	SUPPORT GROUP C (HOURS)	NONLABOR (DOLLARS)
PREPROPOSAL					
BIDDERS' CONFERENCE					
PROPOSAL PREPARATION					
CUSTOMER PRESENTATION					
CONTRACT NEGOTIATION					
OTHER					
TOTAL HOURS					X
TOTAL COST					

FUNDING NEEDED	JAN	FEB	MAR	APR	MAY	JUN	JUL	AUG	SEP	OCT	NOV	DEC
MONTHLY												
CUMULATIVE												

APPROVALS			
	GROUP A DATE	MARKETING MANAGER DATE	VICE-PRESIDENT—FIN. DATE
	GROUP B DATE	DIVISION CONTROLLER DATE	EXEC. VICE-PRESIDENT DATE
	GROUP C DATE	DIVISION MANAGER DATE	PRESIDENT DATE
	PROPOSAL MANAGER DATE	VICE-PRESIDENT—OPERAT. DATE	CHAIRPERSON OF BOARD DATE

Figure 3-3. Typical proposal authorization form.

strained (for instance, some government RFPs), it is vital to spend time with your customer fixing the theme and central focus of your proposal effort. This includes clarifying which dimension(s) of the Triple Constraint should be emphasized.

There are several reasons that theme fixation is important. In the first place, many customers are in fact organizations and the "customer" is comprised of people who view the contemplated undertaking in slightly different ways. It is necessary to understand these subtle differences and either harmonize them or deduce who has the most influence.

In addition, the customer's statement of the problem in the RFP may be imperfect or incomplete. Working with the customer—when this is permitted—can correct this problem as well as demonstrate your organization's competence. In the course of this dialogue with your customer, you will have opportunities to launch trial balloons representing your initial approach to the proposal. You will learn your customer's preference, which will permit you to adjust your thinking to produce a proposal that is more responsive to your customer's prejudices and predilections.

It is important for everybody in your organization to understand the chosen theme so their contributions are consistent with it. This theme will be used throughout your written proposal. It might be technical sophistication and elegance, early delivery, the fact that the unit you propose to furnish is a proven item, or that you have a team ready to put to work on the job. If everyone in your organization understands what the theme is, your proposal should be acceptable to your prospective customer.

Statement of Work

Content

The statement of work (SOW) must describe the job to be done. It should designate any specifications that will be applied. It should identify measurable, tangible, and verifiable acceptance criteria so there is no uncertainty whether the final item is in fact acceptable.

As I said in Chapter 2, when it is not possible to be this precise at the earliest phase of a job, because the final product is not assuredly realistic or clearly attainable, it is important to undertake a two-phase project effort. The first phase, perhaps extending to a customer review, is quoted completely, but the whole job is quoted only approximately, in a nonbinding fashion. The whole job quote represents the proposer's

Project goals must be specific, measurable, and attainable.

best estimate of project requirements. The first-phase quotation, however, is firm (even if a cost reimbursable contract form is used) and includes sufficient effort to construct acceptance criteria for the rest of the job.

Clarification

After it is drafted, the SOW should be reviewed with the customer prior to further work on the proposal itself. Ambiguous words should be avoided and it is desirable to be quantitative, using numbers and dimensions whenever possible.

Make the SOW precise and measurable.

A SOW such as "design and build an amplifier with greater gain than model L" can be ambiguous. The contractor might interpret it as being satisfied by 10 percent greater gain. The customer might expect the new amplifier to have 300 percent greater gain.

It is also important to be sure that extraneous requirements have been eliminated from the SOW. All too often "gold plating"—unnecessary customer or user requests or excessive technological sophistication by your own engineers—creeps in. This should be eliminated if it contributes unnecessary cost or adds too much time to the project schedule.

Plan

A proposal is a bridging step in the overall project process, being involved in both the "define" and "plan" phases of a project. The proposal will contain a plan; additionally, the act of writing the proposal forces the organization to think through and attempt to simulate the entire project.

Simulation

Simulation is used in many situations. Servomechanism engineers, for instance, will simulate on paper, and perhaps further using a digital or analog computer, the performance of a servomechanism before attempting to build even the breadboard. In doing this simulation, they are investigating how the servomechanism might perform if it is built according to certain specifications.

Simulation aids prediction.

Civil or structural engineers will often examine the deformation of a building or a bridge to determine if it will have adequate strength if built according to the design drawings. Similar simulations occur in many other fields, such as aerodynamics, thermodynamics, and optical design. A major purpose of these simulations is to identify any potential problem areas in the prospective system before building it. If a design won't work on paper, it won't work when it's built.

In the case of project planning, where the project plans are a simulation of how the project will be carried out, there are similar reasons for engaging in simulation. It is important to decide how to establish a price for the proposed work. A detailed simulation, that is, a plan, makes it more likely that the proposed or bid price will be sufficient. If the plan has been thoroughly prepared, it will also convince the customer that your organization understands the proposed job, which helps in negotiating your contract. If the project schedule is unrealistic on paper, the project won't be completed on time.

If the schedule doesn't make sense on paper, the project will be completed late.

The Triple Constraint

Because the Triple Constraint is so important to planning and planning is so important to the proposal process, Part 2 of this book is devoted to planning. But it is important to understand that the project plan devised during the proposal process and presented in the proposal is a plan to satisfy the Triple Constraint. You use the work breakdown structure (WBS) to describe your approach to the performance dimension. You use network diagrams or in some cases a bar chart, in which each activity corresponds to a WBS element, to describe your approach to the time dimension. You use a complete cost breakdown for each activity to describe your approach to the money dimension and defend your price.

These three planning elements are best prepared in the order presented. First, the WBS is used to describe those things that will be undertaken to satisfy the performance specification. After that is complete, it is possible to prepare a network diagram for each of these designated elements in the WBS. Initially, each of these items should be estimated in a "natural" time frame; then these activities' logical relationships to each other can be established. If, as usually happens, this produces an unacceptably long program, it is then important to decide which activities will be scheduled to be completed in periods shorter than the "natural" time. That is, some of the project activities must be carried out faster than is desirable. After this has been done, prepare cost estimates for each activity. Note that it is not desirable to prepare the cost estimates prior to determining the time to be allowed for a given activity. These issues will be treated in more detail in later chapters.

Checklists

Checklists are designed to help assure that nothing that will have to be dealt with during the course of the job has been forgotten or omitted from consideration in the proposal.

The plan has three dimensions.

A checklist should contain such items as shown in Appendix 3. These checklists are not exhaustive and may not contain the most significant or most important items. They are meant to suggest the kinds of items that might appear in a checklist.

Take advantage of experience.

The best way to develop a checklist is to create your own over a period of years. One way is simply to enter items on file cards whenever they occur to you during the course of project work. You may later sort out these cards alphabetically, by time phase, or by some other logical method. Having developed a checklist from your own experience, you will perform better on future projects because you will not forget items that are likely to be significant. Thus, you will consider their impact on a project during the proposal phase; they will not emerge as unexpected developments during performance of the project.

On a new proposal or project, think about it critically before consulting your checklist. If you turn to the checklist first, you tend to develop "blinders" to new crucial factors unique to the current job.

Adjustments

Adjustments are often required after a proposal has been partially prepared. Perhaps someone discovers that two departments contributing to the proposal have duplicated their efforts or have made differing assumptions about some significant item. Or perhaps someone discovers new information or corrects some oversight.

When an adjustment is required, all participants must join in deciding how to make it. The proposal manager should not assume this responsibility. Two benefits accrue from participants making the adjustment. First, the experts are considering the problem and presumably making the most sensible adjustment. Second, having contributed to the adjustment, other participants gain a sense of involvement in the decision and tend to perform the job better when the proposal has been converted into a project undertaking.

Approval

As with the initiation of the proposal effort, the conclusion of the proposal requires a managerial action within an organization. There will typically be a sign-off control sheet (Figure 3-4). Normally, each organization has a procedure that

PROPOSAL SUBMISSION APPROVAL	COMPANY PRIVATE

PROPOSAL TITLE	NUMBER

CUSTOMER	CONTRACT FORM

SUMMARY STATEMENT OF WORK

SCHEDULE FOR JOB

COST	FEE	TOTAL BID PRICE

DOCUMENTS, REPORTS, MODELS, ETC., SUBMITTED

SUMMARY OF OUR RISKS

KEY PEOPLE PROMISED

FINANCIAL COMMIITMENTS REQUIRED

WARRANTY

ACCEPTANCE CRITERIA

REMARKS

MARKETING MANAGER	DATE	VICE-PRESIDENT—FIN.	DATE
DIVISION CONTROLLER	DATE	EXEC. VICE-PRESIDENT	DATE
DIVISION MANAGER	DATE	PRESIDENT	DATE
VICE-PRESIDENT—OPERAT.	DATE	CHAIRPERSON	DATE

Figure 3-4. Typical proposal submission approval form.

specifies the signature authority of given managerial levels, and such a procedure indicates which managers or officers must sign the control sheet signifying their approval for proposal submission. The sign-off control sheet must contain a brief description of the Triple Constraint contained in the proposal document being submitted. The sign-off control sheet is retained in the proposing organization's files and is not submitted to the customer.

Plan ahead for the proposal approval.

It is important not to take the approval of senior managers for granted. Therefore, it is important to give these people timely briefings throughout the proposal preparation effort as to the scope of the proposal and the nature of the resources to be committed to the resulting project. Although the proposal authorization document constitutes one such involvement of senior management, it alone will not suffice. The number and frequency of such briefings during proposal preparation depend on the organization, its rules and procedures, and the proposal manager's good judgment.

In the case of proposals to government customers, there is usually a requirement for an officer to sign a certificate of current pricing or similar form. This is intended to assure the government that the proposed price is calculated in conformance with current laws and procurement regulations. Because the signer of this certificate is subject to substantial personal penalties if the certificate is inaccurate, this officer will require a careful review of your proposal SOW and cost estimate.

Submission

The time comes to submit the proposal to the designated recipient, who may require that it bear a postmark by a certain date or be received at a given office by a specified date stamped by a particular time. Such standards are overriding and must be complied with.

Postsubmission

Mailing or delivering the proposal is not the end of a winning proposal effort. At the very least, the winning organization must negotiate a contract with the customer. Sometimes several proposing organizations are deemed qualified and negotiations are carried out with two or more of them prior to selecting the winning contractor.

In many proposal situations, the negotiation phase is preceded by a presentation to the customer. Such a presentation may be elaborate, requiring special graphics and models, and may entail extensive time and effort.

More work—and expense—follows submission.

To recapitulate, the following are steps in the proposal process: Establish the organization's business strategy. Understand the organization's resources. Get the authorization (bid/no bid). Make the preproposal effort. Receive the RFP. Attend the bidder's conference. Fix the theme. Prepare the statement of work. Plan the job. Adjust the proposal. Approve the proposal. Submit the proposal. Present the proposal to the customer. Negotiate the contract. The proposal contains a statement of work, which is the basic project definition. To prepare the proposal, it is necessary to do some, but not all, of the project planning. Much of this planning is commonly included in the proposal.

TYPICAL PROBLEMS

There are both practical and people problems with proposals for engineering projects. The first practical problem encountered in proposing winning engineering projects is attempting to do virtually the entire job during the proposal. That is, in trying to prepare a solid proposal, you spend too much time working through the plan for the project. Sometimes this includes doing preliminary engineering or modeling. You can overcome this by recognizing that risk must be balanced and planning only enough to reduce the project's uncertainty to an acceptable level. A related problem is inadequate project planning in the proposal. The solution here is to keep planning until it becomes too time-consuming. This is obviously a judgmental issue for which personal experience must provide guidance. Another problem is the last-minute rush to complete the proposal in time to submit it. The solution to this problem is to have a proposal preparation schedule and adhere to it.

The principal problem with people is getting the workers who will subsequently perform the work on

continued

the project to contribute to the proposal. The ideal at which to aim—seldom completely achieved in practice—is to have all the people who will be key project participants describe, plan, and estimate the work they will later do. This is the Golden Rule discussed in Chapter 5.

MATERIALS STUDY PROJECT

After being approached by NERESCO, you reviewed their work with SUPROMAC management. SUPROMAC has done prior work for NERESCO and the materials study project is deemed a lead to a still larger effort; so SUPROMAC is anxious to propose on the work, and a proposal is authorized (see Figure 3-5). Your discussions with NERESCO about different initial perceptions of the desired Triple Constraint have led to an agreement to study only two materials, thus conforming to their wish for a six-month project and staying within their budget of $500,000. You have also agreed on which two materials will be studied. But they want a theoretical study included to reduce the risk resulting from omitting the study of the other four materials you felt should be studied.

Thus, you can now write the proposal. The SOW will name the two materials, the specific tests to be performed, and the kind and amount of data to be delivered. It will also contain a detailed schedule of work and cost breakdown.

PROPOSAL AUTHORIZATION		NUMBER *1728*	REVISION *ORIG.*

TITLE *MATERIALS STUDY*

<table>
<tr><td rowspan="11">JOB</td><td colspan="4">PERFORMANCE REQUIRED
STUDY 2 MATERIALS</td></tr>
<tr><td colspan="2">ESTIMATED STARTING DATE
JAN 84</td><td colspan="2">ESTIMATED DURATION *6 MONTHS*</td></tr>
<tr><td colspan="2">ESTIMATED BID PRICE
$500,000.00</td><td colspan="2">ESTIMATED SUBCONTRACT TO OTHERS
$ *$5000.*   % *1%*</td></tr>
<tr><td>IS JOB FUNDED?
YES</td><td colspan="3">WHAT IS FOLLOW-ON POTENTIAL?
VERY LARGE ($5M?)</td></tr>
<tr><td colspan="4">ESTIMATED NEED FOR CAPITAL AND FACILITY EXPENSE IF JOB IS OBTAINED
USE EXISTING EQUIPMENT & SPACE</td></tr>
<tr><td colspan="4">CUSTOMER ORGANIZATION
NERESCO</td></tr>
<tr><td colspan="4">KEY CUSTOMER PERSONNEL
KIM DRIVER</td></tr>
<tr><td>CONTRACT FORM
FP</td><td colspan="2">SPECIAL CONSIDERATIONS
—</td><td>SECURITY CLASS
UNCL</td></tr>
</table>

<table>
<tr><td rowspan="5">COMPETITION</td><td>COMPETITORS *UNCERTAIN — PROBABLY SOLE SOURCE*</td></tr>
<tr><td>COMPETITORS' STRENGTHS
UNKNOWN</td></tr>
<tr><td>SIGNIFICANCE TO COMPETITORS IF THEY LOSE
UNKNOWN</td></tr>
<tr><td>OTHER COMPETITOR WORK FOR CUSTOMER
UNKNOWN</td></tr>
<tr><td>OUR UNIQUE ADVANTAGES
LOT'S OF DISCUSSION, CAN DO IT, MAYBE SOLE.</td></tr>
</table>

PROPOSAL section:

WHAT IS TO BE SUBMITTED? *PROPOSAL*

DUE DATE *15 NOV 83*	PROPOSAL COST (DETAIL BELOW)
PROPOSAL MANAGER *MEL CHASE*	OTHER KEY PROPOSAL PERSONNEL *CHERYL WILLIAMS*

ACTIVITY / EFFORT	PROJECT DEPT. (HOURS)	SUPPORT GROUP A (HOURS)	SUPPORT GROUP B (HOURS)	SUPPORT GROUP C (HOURS)	NONLABOR (DOLLARS)
PREPROPOSAL	*DONE*				
BIDDERS' CONFERENCE	—				
PROPOSAL PREPARATION	*20*	*8*			
CUSTOMER PRESENTATION	*4*	*4*			
CONTRACT NEGOTIATION	*8*				
OTHER					
TOTAL HOURS	*32*	*12*			
TOTAL COST	*800.*	*300.*			

FUNDING NEEDED	JAN	FEB	MAR	APR	MAY	JUN	JUL	AUG	SEP	OCT	NOV	DEC
MONTHLY											*0.7*	*0.4*
CUMULATIVE											*0.7*	*1.1*

APPROVALS:

GROUP A	DATE	MARKETING MANAGER	DATE	VICE-PRESIDENT—FIN.	DATE
GROUP B	DATE	DIVISION CONTROLLER	DATE	EXEC. VICE-PRESIDENT	DATE
GROUP C	DATE	DIVISION MANAGER	DATE	PRESIDENT	DATE
PROPOSAL MANAGER	DATE	VICE-PRESIDENT—OPERAT.	DATE	CHAIRPERSON OF BOARD	DATE

Figure 3-5. Typical proposal authorization form.

HIGHLIGHTS   *Organizations must filter out losing projects.*

Winning projects arise from good proposals, thoughtfully initiated to be consistent with the organization's goals.

Four issues involved in the decision to bid on a proposal opportunity are the nature of the requirement, the value of the project, the organization's response ability, and its ability to win.

The proposal defines what the project will accomplish.

A checklist may help you avoid overlooking required work.

The proposal process includes authorization, theme selection, SOW preparation, plan development, adjustment, approval, submission, and postsubmission follow-up.

Project proposals describe the Triple Constraint with a work breakdown structure, activity network diagram or bar chart, and cost estimate for each activity, which then serve as a project plan.

FURTHER READING

V. G. Hajek. *Management of Engineering Projects.* New York: McGraw-Hill, 1977.
> *Chapter 4 contains a good description of the content of a proposal for an aerospace system project.*

R. J. Loring and H. Kerzner. *Proposal Preparation and Management Handbook.* New York: Van Nostrand Reinhold, 1982.
> *This handbook covers the preparation and submission of proposals.*

P. W. Metzger. *Managing a Programming Project*, 2nd ed. Englewood Cliffs, NJ: Prentice-Hall, 1981.
> *Pages 174–176 enumerate ten points about proposals for computer software projects.*

Negotiations and Contracts

Negotiations between the customer and the contractor convert the final definition of work into a contract. This chapter emphasizes the importance of appropriate contract forms and the necessity to start the proposal process with attention to this issue. It also discusses the special case of projects involving a foreign customer.

NEGOTIATING THE CONTRACT

Preproposal Negotiations

Contract negotiation really begins in the proposal phase because the expected contract form must be consistent with the job to be undertaken. If there is any reason to believe the customer will require a fixed price contract (Table 4-1), for instance, and the job calls for a major technological advance you are not certain you can achieve, it would not be prudent to continue in the preproposal and proposal effort. Hence, one objective of preproposal activity and discussions with a customer prior to major proposal expense is to assure that the contract form they intend to issue is consistent with the contract form your company or organization is willing to negotiate considering the work to be undertaken. In addition, negotiations are designed further to improve the likelihood of the customer organization and contracting organization having the same perception of the job.

Make a proposal only if the contract will be acceptable.

TABLE 4-1.   Common Contractual Forms.

Abbreviation	Definition
FFP	Firm fixed price—the price and fee are predetermined and do not depend on cost
FP	Fixed price—same as FFP
CPFF	Cost plus fixed fee—the customer agrees to reimburse the contractor's actual costs, regardless of amount, and pay a negotiated fixed fee independent of the actual costs
CPIF	Cost plus incentive fee—similar to CPFF except the fee is not preset or fixed but rather depends on some specified incentive
T AND M	Time and material—the customer agrees to pay the contractor for all time and material used on the project, including a fee as a percentage of all project costs

Customer and Contractor Perceptions

The perceptions of both contracting parties must be harmonious. It is always possible for an RFP or the responding proposal not to be completely clear. If the preproposal process did not remove these potential misperceptions, the negotiation process offers the last opportunity to do so. Specifically, the final deliverables—software, firmware, and/or hardware—must be well defined, and the criteria for measuring or judging acceptance and completion must be straightforward.

Contractual Forms

At the simplest level, there are no contracts. This is typical of self-financed efforts such as a project to paint your own house. This situation also prevails in efforts supported by your own organization, such as development of a new product's production line for use within your own company.

When one organization enters into a contract with another organization, there is a variety of possible contractual forms (see Table 4-1). In the first of these, very common in commercial situations, the contract is a so-called fixed price (FP) or firm fixed price (FFP) contract. This has the lowest financial risk to the customer because the maximum financial obligation is specified; conversely, the FP form has the highest financial risk for the contracting organization but offers the highest potential reward if the estimated costs can be underrun. There are variations on this in which the price allows for escalation or redetermination due to some set of factors, such as inflation. Or there may be an FP contract with an incentive fee based on some performance aspect, perhaps early delivery.

Another class of contractual arrangements is those in some way cost reimbursable. Here the customer bears an obligation to reimburse the contractor for all costs incurred; so the customer has a high financial risk and the contractor has a correspondingly low risk. Typical contracts of this sort are cost plus a fixed fee (CPFF) or cost plus an incentive fee (CPIF). Time and material (T and M) contracts are also a form of cost reimbursable contract.

The contract form puts risk on either the customer or the contractor.

Contracts entered into with the U.S. Department of Defense are governed by complex regulations, the Defense Acquisition Regulations (DAR). In the case of the FP contracts, the government is never obligated to pay more than the specified amount. However, if the contractor performs very well and manages to underrun the cost budget substantially, the government has the right to reduce the amount paid to the contractor below the price specified in the contract. Thus, FP contracts are a one-way street to the government's advantage.

An FP contract is preferable when there is virtually no uncertainty about your ability to satisfy the performance specification on schedule and for, at most, your estimated costs. Such situations typically occur when you have done essentially the identical or very similar project work previously and you have the appropriate human and physical resources available. If these conditions are not satisfied, the FP contract is a financial gamble and should be undertaken only if the risk is acceptable and the prospective reward is commensurate with it.

Negotiations

In a typical negotiation, the customer attempts to increase the performance specification while reducing the schedule and budgeted cost. If it is a competitive solicitation, the customer will often play off one prospective contractor against another to try to maximize his or her apparent benefit. Therefore, one should expect the customer to behave in this way.

Expect the negotiated contract price to be less than your proposed price.

In addition, negotiators should understand clearly how far, if at all, their management is prepared to deviate from the terms and conditions offered in the submitted proposal it approved. Second, good planning aids negotiators in that there is a complete work breakdown structure, with an attendant activity schedule and cost estimate for each element of it. These help negotiators understand the job being negotiated and can usually help explain and/or defend it to the prospective customer.

In preparing for negotiations, it is frequently desirable for your organization to conduct a trial run with someone or some group playing the customer. In short, be well prepared and know your minimum acceptable position. Also, because only one negotiating team member should talk at once, you can rehearse who will respond to particular issues if they are raised.

After two dimensions of the Triple Constraint have been specified, the third can be determined.

During actual negotiations, it is usually to your advantage first to define the job (the detailed statement of work, specifications, and test criteria) and the schedule. After that, you can negotiate the exact contract form, including any detailed terms and conditions, and the final price. The effective negotiator always "horse trades" and never makes a unilateral concession.

Nevertheless, there is often give and take in the negotiation process, and some changes may be agreed to. Whenever there is a change in one element of the Triple Constraint, there must be changes in other elements. For instance, a customer may offer to provide customer furnished equipment (CFE) to reduce the expected cost of some activity within the proposed project. In the case of the government, this is called government furnished equipment (GFE). When this occurs, language on the performance axis must be changed to indicate the performance specifications the CFE must meet and the schedule of CFE provision must be stipulated. When both these things have been accomplished, it is possible to agree to substitute the CFE for contractor procured items and offer a reduction in the proposed schedule or budget.

A negotiated change in one dimension of the Triple Constraint should be accompanied by changes in the other dimensions.

When the negotiations are concluded, the binding direction on the project is that specified in the contract. Only contract change notices suitably signed and agreed to by both parties can permit changes. Such changes should not occur outside the contracting mechanism, despite agreements reached by members of the contractor's and customer's support teams—for instance, when two engineers meet and agree that a new microcircuit would be a desirable item to include. Verbal redirection, which can easily occur because of the many individuals involved, is not binding until the contract is amended, as shown in Figure 4-1. Renegotiations such as these require the same amount of planning and preparation as the original negotiations. It is not uncommon to have several renegotiations during a long or complex project.

Legal Aspects

There is a myriad of legal aspects in project contracts. Discussions with suitable people in your organization are often helpful.

In the case of the U.S. government contracts, there is a host of special regulations. A typical government contract will include the following:

1. Customer's name and address and specification of authorized signatures

2. Contractor's name and address and specification of authorized signatures

3. Statement of supplies (items), services, and prices (statement of work)

4. Preservation, packaging, and packing instructions

5. Delivery or performance period

6. Inspection and acceptance terms

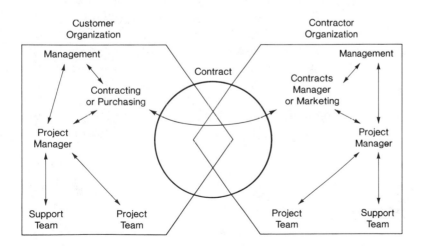

Figure 4-1.   Interpretation of the Triple Constraint must be controlled through the contract.

7. Contract administration data

8. Special provisions (funding limitations or customer furnished equipment)

9. General provisions (reference to Defense Acquisition Regulations or overtime payment terms)

10. Patent terms, conditions, and ownership rights (if not covered in items 8 and 9)

11. List of required documentation

Misrepresentation of costs is illegal and can produce serious consequences. There can be other problems with government contracts. A semiconductor manufacturer recently paid $1.75 million in criminal and civil penalties because the company violated federal regulations concerning testing of electronic chips used in military equipment (*New York Times*, March 7, 1984, p. D1 and March 11, 1984, p. 14F). In addition, in the case of fixed price contracts, failure to deliver can expose the contractor to very serious cost penalties far beyond absorbing the cost of its own effort. Such penalties may not only include termination but may obligate the contractor to pay the costs of a substitute the government hires to deliver after the contract is in default.

If any patents are obtained in the course of the project, the government may own them. In addition, there are frequently highly proscribed procedures specifying how much of the work throughout the project must be performed. Costs are frequently subject to audit when the job is completed, and the amount finally paid to the contractor is frequently reduced.

Understand the legal implications.

In the case of commercial contracts, restrictions are far less one-sided. Force majeure (that is, an act of God) may permit a contractor to escape any penalty for failure to deliver on time. In addition, the suitably drawn contract will specify that disputes are to be settled by arbitration rather than by the courts. Arbitration is far simpler, quicker, and less costly to both parties. Nevertheless, antitrust regulations and many other laws limit the kind of commercial arrangements into which two companies may enter.

INTERNATIONAL PROJECTS

International projects are not fundamentally different from domestic projects. In practice, however, the travel lure of a remote and apparently salubrious destination seems to affect human judgment. Furthermore, international projects expose a contracting organization to special problems related to language, currency, and unfamiliar business practices.

Remoteness

The other party in an international project is typically located in some remote area of the world. The far-off hills often look green. When two project opportunities are available to a company or organization at the same time, the one that originates in Zurich will receive far more attention than the one that originates in Cleveland, and this is independent of the intrinsic merits of the project opportunity. This is the "travel lure," and it is a problem you must recognize and identify for what it is.

Recognize the travel lure.

There are occasions, however, when you should undertake a project involving a foreign customer or partner. When this is the case, it is desirable to get more information about the foreign organization or the business in which it is engaged. If your organization has an office in the country, that is the best source of information. The foreign organization's country may have an information office in the United States, which may be able to provide useful information. The United States embassy or trade centers in the foreign country may have information, as may the U.S. Department of Commerce or the World Bank.

Having gathered all the information you can without leaving the United States, you may then have to succumb to the travel lure and go to the foreign country to discuss the project or negotiate. It is generally prudent to go several days before the initial meeting to compensate for jet lag.

Business Practices

The business practices governing engineering project performance as well as negotiations and discussions leading to the project will typically be those dictated by the customer's country. For instance, Japanese tend to discuss a project at great length to try to achieve consensus. When this has been done, a handshake or verbal agreement is binding.

Staffing requirements will frequently require that some of the work be reserved for the customer's nationals. Sometimes nationals of a third country are stipulated or prohibited, for a variety of reasons. Permits or "red tape" enter into business practices, and these will often absorb far more time than expected. In some cases, export or import controls can impose severe restrictions; so it is always important to understand these issues early.

Laws and Regulations

There are numerous laws and regulations, unfamiliar to the vast majority, that apply to international business. Limitations on the export of technology from the United States make headlines periodically when someone or some company is charged with a violation. Such violations typically involve shipment (or indirect reshipment) of some restricted high-technology product to the Soviet Union.

In other situations, even more arcane restrictions, such as the Export Administration Act, may snag the unwary engineering project manager. In fact, a United States congressman, trying to help a high-technology company in his district, is reported (*Wall Street Journal*, February 22, 1984, p. 5) to have broken a law prohibiting U.S. citizens or companies from aiding the Arab boycott of businesses tied to Israel merely by writing a letter to someone in Kuwait. Thus, if you are contemplating an international engineering project, be sure to have competent legal guidance on what you can and cannot do.

Language

Be prepared for unfamiliar customs, language, and currency.

Fortunately for Americans and people from other English-speaking nations, most contracts are written in English. In some cases, however, a customer's language will prevail, and the contract will be written in it. If you are using a translation, it is prudent to have several prepared because the original host language document is the controlling document and interpreters are likely to interpret it differently. If you find wording differences, you can explore their significance with a language expert.

Price

As with language, the customer's currency may be the stipulated medium of financial settlement. Because currency rates fluctuate, it may be desirable to insure your company against these fluctuations. This is done with a hedge contract. Alternatively, you may use letters of credit to satisfy the future payment obligations. In any event, your price must allow for the extra expense of doing business at a great distance.

TYPICAL PROBLEMS

The most common negotiation problem is deciding how to cope with the inevitable price squeeze the customer will try to inflict. Several things help:

1. A good plan, well explained

2. A clear understanding of where you have inserted negotiating "cushion" or "fat" (as distinct from contingency, which is discussed in Chapter 11) and a negotiating plan on how to horse trade it

3. Management guidance, or clearance, on how much you can give up

4. A reputation for having met prior commitments

Finally, if you are forced to surrender schedule or price, be sure to try to alter the performance specifications or obtain CFE.

MATERIALS STUDY PROJECT

Preproposal discussions between SUPROMAC and NERESCO have narrowed the scope of negotiations considerably. Because SUPROMAC believes this project will lead to more work later, it is not interested in taking a tough and unbending posture. Nevertheless, SUPROMAC might first try to justify a price in excess of $500,000 because of the theoretical study NERESCO substituted for the four materials omitted from the study.

continued

> On the presumption that an FP contract is jointly agreed to, SUPROMAC can attempt to gain favorable detailed terms and conditions. These might include progress payments and a degree of informality in the required reports.

HIGHLIGHTS

A proposal should be made only if a reasonable contract can be negotiated.

There are several contractual forms, including fixed price, firm fixed price, cost plus fixed fee, cost plus incentive fee, and time and material.

Both parties must be prepared to make concessions during negotiations.

Government contracts include numerous special regulations.

International projects introduce special problems, such as unfamiliar business practices, laws and regulations, language, currency, and distance.

FURTHER READING

R. H. Clough and G. A. Sears. *Construction Project Management,* 2nd ed. New York: Wiley-Interscience, 1979.
> *Chapter 1 discusses contracting and negotiating issues in the construction industry.*

R. Fisher and W. Ury. *Getting to Yes.* Harmondsworth, Middlesex, England: Penguin, 1983.
> *This is a very readable, short book on techniques for negotiating an agreement without giving in.*

J. L. Graham and R. A. Herberger, Jr. "Negotiators Abroad—Don't Shoot From the Hip." *Harvard Business Review* (July-August 1983), pp. 160–168.
> *This article offers practical advice on international negotiations.*

V. G. Hajek. *Management of Engineering Projects.* New York: McGraw-Hill, 1977.

    Chapters 7–10 provide an extensive and thorough treatment of contracts, negotiations, and legal issues.

V. Maieli. "Sowing the Seed of Project Cost Overruns." *Management Review* (August 1972), pp. 7–14.

    This is a general discussion of the causes of cost overruns, with very useful pointers on traps to avoid during negotiations.

J. Main. "How to be a Better Negotiator." *Fortune* (September 19, 1983), p. 141.

    This short article touches on some of the points in the Fisher and Ury book.

M. Silverman. *Project Management—A Short Course for Professionals.* New York: Wiley Professional Development Programs, 1976.

    Chapter 4 provides a fairly extensive coverage of negotiations and contract forms.

Part **2**

PLANNING AN ENGINEERING PROJECT

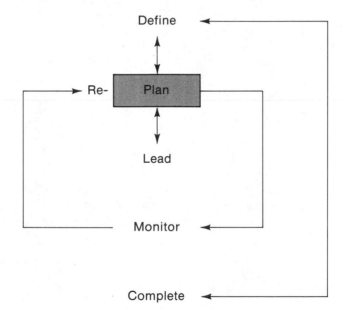

5

Why and How to Plan an Engineering Project

The planning activity for the management of an engineering project is crucial. Plans are the simulation of a project, comprising the written description of how the Triple Constraint will be satisfied. Therefore, project plans are really three plans: one for the performance dimension (the work breakdown structure), one for the schedule dimension (preferably a network diagram but occasionally a milestone listing or bar chart), and one for the cost dimension (a financial estimate). This chapter covers plans in general, reiterates the need for plans, describes how these three kinds of plans are made, reviews several planning issues, and discusses "the Plan," a document or series of documents embodying the project's planning agreements.

PLANNING

In broadest generality, plans depend on three factors:

1. Knowing where you are now (or will be when whatever is being planned for will start)

2. Knowing where you want to get

3. Defining which way you will get from where you are to where you want to be

These factors are illustrated in Figure 5-1. The old saying, "When you don't know where you want to go, any road will

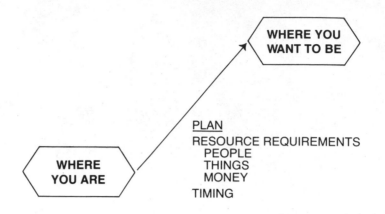

Figure 5-1. Planning.
Source: "Basic Management Skills for Engineers and Scientists." Lecture notes by Milton D. Rosenau, Jr. Copyright © 1982 by the University of Southern California. Reprinted by permission.

get you there," is true; you can have a plan only if you have a destination in mind.

Many company sponsored engineering projects are initiated because of the organization's long-range plan. Thus, plans are frequently hierarchical, with short-range plans established within the context of long-term plans. For instance, project task plans are components of the overall project plan. When the long-range plan covers five or ten years, changes obviously occur, priorities must be altered, and projects are added or canceled in response to the dynamic environment.

THE NEED FOR PLANS

Plans aid coordination and communication, provide a basis for control, are often required to satisfy requirements, and help avoid problems.

Coordination and Communication

Most engineering projects involve more than one person. Typically, a technical expert is asked to perform in the area of his or her expertise. For instance, an expert on electronic circuit design works on the electronic circuit design task, not on the optical design task for an electro-optical system. The project plan is a way to inform everyone on the project what is expected of him and her and what others will be doing. Plans are a vehicle to delegate portions of the Triple Constraint down to the lowest (task or subtask) reporting level. If the people responsible for these tasks also participate in making the plans, they will have an added impetus to adhere to them. Thus, there is a Golden Rule for planning:

Let others plan their work.

Get the persons who will do the work to plan the work.

- They should know more about it than anyone else.
- It's *their* task, not *yours*.

Your project plans matter. Even if your project can be performed in your office, other people in the organization (for instance, your boss) will want to know where your project is headed, what you are doing, and for how long you will be doing it. Thus, project plans constitute an important communication and coordination document and may motivate people to perform better.

Basis for Control

Plans are also the basis of your project controls (see Figure 5-2). It is a characteristic of projects that they do not go in accordance with plan. What you do not know when you start is where and how your project will go off plan. Deviations from plan, detected by monitoring progress, constitute your early warning signal during project performance that there are problems to be resolved. That should cause replanning to occur.

Projects do not follow their plan.

When a new navigation system is being tested, it could be installed in a flight test airplane. The airplane is instructed to fly a precalibrated course with several fixed reference check points. During the course of the flight test, observations of the navigation system signals are compared to the expected signals at each check point. When deviations are observed, adjustments must be made. The same is true of a project deviating from the plan to achieve its Triple Constraint.

Plans are a detailed description, formulated before the project is carried out, for accomplishing its various aspects. Deviations may indicate that the project will not reach its intended destination.

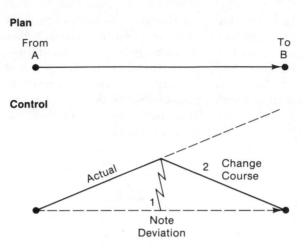

Figure 5-2.   Plans are the basis for control.

Source: "Project Management." Lecture notes by Milton D. Rosenau, Jr. Copyright © 1981 by the Association for Media-based Continuing Education for Engineers, Inc. (AMCEE). Reprinted by permission.

There are plans for all three dimensions of the Triple Constraint.

Figure 5-3 illustrates how all the techniques covered in Part 2 are related. The work breakdown structure (WBS) leads to a network diagram. Each activity or task on the network is then time-phased and checked for resource allocation to be certain it is consistent. When this is done, it is possible (although not necessary) to construct a schedule bar chart indicating the time frame for each activity or task of the entire project. The costs for these tasks are then estimated and, when work is being performed on them, costs are reported back by specific task in each expense category. Part 4 discusses how this detailed variance information can be used to help manage the project.

Requirement Satisfaction

Plans are sometimes created merely to satisfy requirements imposed by others, perhaps a customer or your boss. In such a situation, plans are often created under duress rather than because they are perceived to be valuable, even essential, in achieving project objectives.

Plans so created are frequently not followed. All too often they are generated and then discarded because they were prepared only to meet the requirement to prepare a plan. When there is such a requirement and the plans are prepared

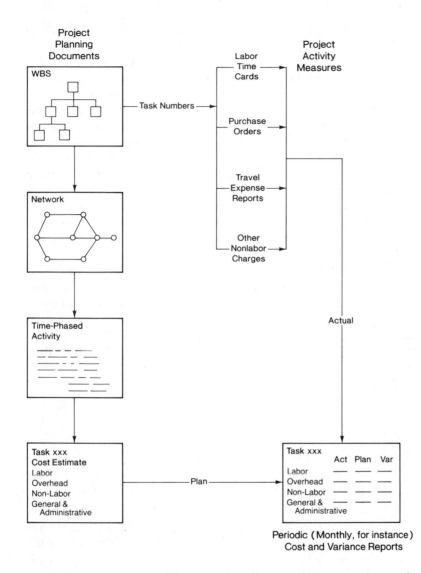

Figure 5-3. Overview of how project planning documents are used in project monitoring phase.

slavishly rather than thoughtfully, it is a waste of time for the preparer and the reader.

Problem Avoidance

Project management is sometimes a race with disaster. This is frequently the case for less experienced project managers. All too often the last crisis has scarcely been resolved before the current one begins, and then the project manager is too busy to anticipate and try to head off the next one.

Planning is crucial; a good plan is essential.

A good plan helps you avoid problems during project performance (but plans cannot prevent problems). Consider the following example of a schedule and cost problem. Your project requires a final report. You assume it will require sixty pages and twenty figures, and the technical documentation group, which will prepare the report, agrees to do it in one week for $1,200. When you later ask them to prepare a one-hundred-twenty-page documentation report with forty figures, you will be told you will get it two weeks later and it will cost $2,400. Obviously, you cannot "plan" for a two-hundred-page documentation report with one hundred figures because the price in your proposal may be too high for you to win the job. You must make the best plans and estimates you possibly can and then try to adhere to them. For instance, as you get to the final phase of the project and must prepare the report, assign writing to the participants in such a way that they all clearly understand the planning goal. Constrain the writing efforts to adhere to plans. If you do this, the report should be approximately sixty pages and twenty figures.

PLANNING ISSUES

Uncertainty and Risk

Plans relate to future events. That is, your plans are a simulation of how things will occur in the future. There are necessarily uncertainties about the future, some of which may be somewhat predictable and thus partially controllable, but many of which are unpredictable.

Use checklists to reduce uncertainties.

You can reduce (but not eliminate) these predictable uncertainties by using checklists, thoroughly discussing the plans with experts, and involving your entire team. Nevertheless, uncertainties will remain because there are always unpredictable factors when you are doing something new. Allowance for these unknowns can be made by inserting contingency in your plans, but the unknowns cannot be eliminated. For instance, thorough plans cannot prevent bad weather or strikes from delaying a construction project or eliminate cost changes due to currency rate fluctuations. Plans can be no better than your present understanding. If you have done something similar before, you can plan it better than if it is entirely new to you and your team. For instance, previous experience with an organic polymer formulation is not terribly helpful for planning a materials study project.

Assumptions, such as which people will be able to work on your project, are involved in planning. The plan for the materials study project looks a lot different when a senior mechanical engineer will do the work than when a junior electrical or chemical engineer will. Because assumptions are involved in your planning, it is important to include contingencies, which I discuss in Chapter 11. Good plans are quantitative rather than qualitative and as precise as possible.

A Choice Between Options

In preparing plans, as in carrying out project work, you are frequently confronted with options. Your plan may be considered the record of your choices between these options and will normally depend on how much risk you are willing to take or how much contingency allowance is included in your plans. Figure 5-4 illustrates this kind of choice. It shows two possible activity sequences. Which you choose depends on what is important to you, in this case, a short schedule (option 2) or the higher assurance of instrument accuracy (option 1). Another possibility (option 3), with two activities being performed simultaneously (in "parallel," as it is sometimes called), requires two people (one to verify, one to collect); the other two options require only one person.

Project participants will frequently present a plan that seems absurd to you. It may in fact be absurd. But perhaps the person who prepared it is simply emphasizing activities that you are not stressing.

Heed others' plans and suggestions.

A common engineering project activity, ordering required materials, illustrates this problem. Sensible choices are to order these materials as early as possible (to be certain they are available when required) or as late as possible (to reduce the possibility of having to change selection or to help your organization's cash flow). It is important to discuss the perceptions of everyone involved in the undertaking.

Or imagine your boss asks you at 9 A.M. to join him or her on a 5 P.M. transcontinental flight to attend an important meeting. You concur and agree to meet at the airport. This arrangement allows you to drive past your house en route to

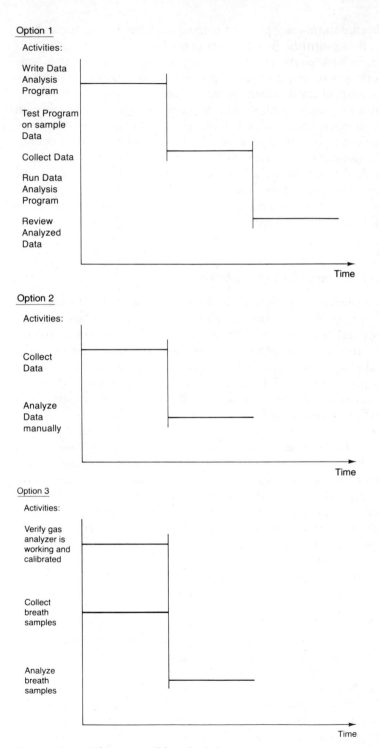

Figure 5-4. Three possible schedule sequences.

the airport and pack your suitcase. In this simple example, your plan might be as shown in Figure 5-5. For this case, the sequence of activities A, B, and C may not seem to matter. And it does not in terms of your time. But it is desirable to start your secretary's assignment as early as possible so he or she can perform activity D while you do activities B and C. Thus, you should perform activity A first, but you can do C before B (or vice versa) without any schedule delay.

Activities:

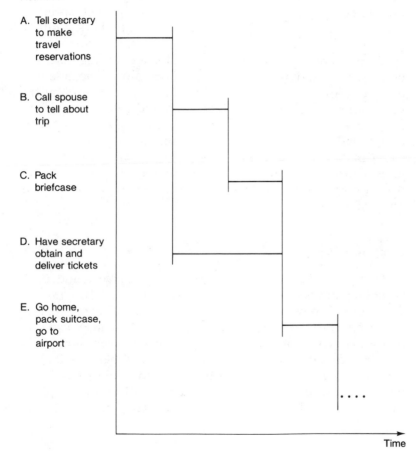

A. Tell secretary
to make
travel
reservations

B. Call spouse
to tell about
trip

C. Pack
briefcase

D. Have secretary
obtain and
deliver tickets

E. Go home,
pack suitcase,
go to
airport

Time

Figure 5-5.   Option for trip preparation.

Hazards

There are innumerable hazards in preparing engineering project plans. One follows from the preceding discussion. In

Planning their own work can motivate people.

an attempt to gain time in the early phases of a project or because you are addicted to your own ideas, you may tend to do much of the planning yourself. You should avoid doing so for the same reason you do not like to be told to carry out somebody else's plan: It is demotivating. In fact, it is important to involve the people who will actually be doing the work so they plan as much of their work as possible. Again, this is the Golden Rule.

In addition, poor planning frequently occurs. Other than sheer laziness, the basis of almost all poor planning is a misunderstanding of the Triple Constraint point. Taking the time to create plans allows you to identify your perception of the Triple Constraint point and shows if and where it differs from somebody else's.

Occasionally, a tool commonly called a planning matrix is used. It lists activities to be carried out along one side of a piece of paper and designates involved personnel along the perpendicular side. Where these rows and columns intersect at a check mark, the designated personnel are involved in the designated activity. This kind of document may be helpful to some managers, but it is a misnomer to call it a planning matrix rather than an involvement matrix. To put it another way, a planning matrix may be a helpful document, but it is not a plan.

Currentness

Keep everyone current on revisions.

Once you have decided to plan your project and have issued the plans, people should take them seriously. They can do so only if they know the plans are current. Therefore, it is very important to know who has copies of them. When you revise plans, be absolutely certain to provide revisions to all people who have copies of previous plans. When you do this conscientiously, everyone involved in your project will know that you take planning seriously. They will know the plans they have reliably indicate the project intention.

"THE PLAN"

In many projects, there is a book called "the Plan" or "the Project Plan." It may be one or a series of thick notebooks.

Issues Addressed

"The Plan" addresses many topics. It frequently describes what is to be delivered as a result of the project and any specifications for those items. If there will be acceptance tests, it will describe them in detail. In general, the purpose of this kind of plan is to describe what is to done, by when, and for how much, in other words, to expand upon the Triple Constraint.

Topics Covered

Typically, "the Plan" will cover many of the following topics:

1. Project summary
2. Project requirements
3. Milestones
4. Work breakdown structure
5. Network diagram of the activities with schedule dates
6. Budget for all activities
7. Project management and organization charts
8. Interface definitions, including facility support
9. Logistic support
10. Acceptance plan
11. Standards for property control and security
12. Customer organization contact points, if relevant
13. Nature of project reviews

Activation

Project plans require activation. First, obtain whatever higher level approvals are required, including those of the customer. Second, disseminate "the Plan" to all involved personnel. In very large projects, dissemination may require a chart room in which the walls are covered with charts displaying the plans for and status of various activities, including financial progress and resource allocation. Chart rooms are not required for smaller projects.

Project plans may vary from fairly simple one-page statements to records with overwhelmingly intricate levels of de-

Spend no more time planning than you would spend correcting problems resulting from having no plan.

tail. There is an appropriate level for each project undertaking. There is no magic formula that establishes the right level of detail; in general, never spend more time planning than it would take to correct any problems encountered because planning had not been undertaken. That is, a basic purpose of planning is to avoid problems.

TYPICAL PROBLEMS

There are three pervasive problems with planning. First, taking enough time to plan is costly. There is an old saying about this: "We don't have enough time to plan now; but we'll have lots of time to fix it up later." In fact, a little inexpensive planning early usually avoids a lot of very costly fixing later. This is analogous to testing engineering breadboards and prototype models—a lot of early testing is much cheaper than trying to remove production problems later. It is difficult to decide how much planning is appropriate, but the inexperienced project manager usually does far too little.

Second, plans are frequently ignored because they are perceived as an irrelevant requirement of management. The solution is obvious: Write meaningful plans you intend to follow and keep current, and be sure everyone understands you have done so.

Third, a separate plan is required for each of the three dimensions of the Triple Constraint. These three plans must be integrated and consistent, and must not—as is frequently the case—be prepared by separate groups of specialists.

MATERIALS STUDY PROJECT

As project manager for the SUPROMAC materials study project, you have to decide what kind of planning to do. The next four chapters illustrate the work breakdown structure, schedule, and task cost estimates you

continued

should create and use. Because it is not a complex project, you can reasonably omit "the Plan" document and a chart room.

However, the research department will be responsible for the theoretical studies that were substituted for testing more materials. Thus, you will arrange to involve them in the detailed planning of their work. (Clearly, there should have been some discussion with them earlier, when this idea arose in negotiations with NERESCO.)

HIGHLIGHTS

Plans are used for the following purposes: simulate how the project will be carried out; write the proposal; negotiate the contract; coordinate and communicate; increase motivation of participants; control the project; satisfy requirements; avoid problems; and record the choice between options.

Plans delegate portions of the Triple Constraint to the lowest reporting level.

Plans help keep engineering projects on course.

If formulated only to meet someone's requirements for them, plans are virtually useless.

Everyone involved must receive every plan revision.

Planning notebooks, a series of planning notebooks, or a chart room may be needed.

Never spend more time on a plan than would be required to correct problems resulting from a lack of a plan.

Plans reflect the balance between risk and contingency for both controllable and uncontrollable future events.

FURTHER READING

R. D. Archibald. *Managing High-Technology Programs and Projects*. New York: Wiley-Interscience, 1976.
Chapter 7, sections 1–4, briefly covers many of the issues in this chapter and contains a complete enumeration of applicable techniques.

P. W. Metzger. *Managing a Programming Project,* 2nd ed. Englewood Cliffs, NJ: Prentice-Hall, 1981.

The portion of Chapter 2, pages 21–49, on project planning is especially useful for computer projects but has general utility as well.

M. D. Rosenau, Jr., and M. Lewin. *Software Project Management: Step by Step.* Belmont, CA: Lifetime Learning Publications, 1984.

This book provides a thorough treatment of all the items for which planning is required in a computer software project.

M. Spinner. *Elements of Project Management: Plan, Schedule, and Control.* Englewood Cliffs, NJ: Prentice-Hall, 1981.

Chapter 5, pages 62–88, discusses project planning, especially project costs.

The Work Breakdown Structure

The goal of a performance dimension plan is to be sure that everything required to satisfy the entire performance specification is done. This chapter deals with planning for the performance dimension of the Triple Constraint. The statement of work is a useful aid, but the principal tool discussed is the work breakdown structure.

STATEMENT OF WORK

The statement of work (SOW) is that portion of the contract that explicitly enumerates what the contracting organization will do for and deliver to the customer. In engineering projects within your organization, the SOW may be contained in a memo or work authorization rather than in a contract, but it should still contain a specific, measurable, and attainable goal. The SOW must always contain a list of all deliverables and should be accompanied by a project schedule and budget to be meaningful. Thus, a plan for the performance dimension of the Triple Constraint is primarily a listing of every activity that must be performed and every result that must be obtained. The SOW frequently contains explicit acceptance criteria and test specifications.

WORK BREAKDOWN STRUCTURE

Purpose

Use a WBS to subdivide projects into tasks.

The work breakdown structure (WBS) is a convenient method for dividing an engineering project into small work packages, tasks, or activities. A WBS reduces the likelihood of something dropping through a crack. To put it another way, a WBS is intended to assure that all the required engineering project activities are logically identified and related.

Figure 6-1 is a WBS, in this case, for a photovoltaic solar power system. There is no magic formula for constructing a WBS. Figure 6-1 shows two levels of detail, but there is no standard number of levels to use. In general, probably at least three or four should be shown, but it might sometimes be appropriate to show five or ten or even more. The breakdown might occur using earlier or later activities, particular organizational involvements, or almost anything that makes reasonable sense. (Note that Figure 6-1 includes the hardware for a photovoltaic solar power system, but the WBS does not list a shipping container, the installation and operation instructions, a warranty document, or a user training manual,

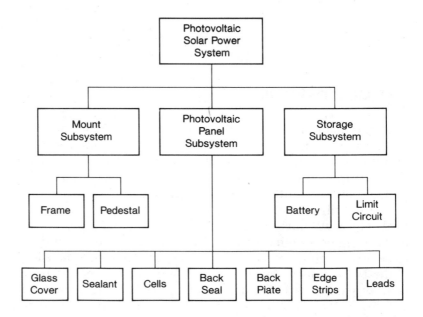

Figure 6-1. Work breakdown structure.

some of which might reasonably be a part of such an engineering project.)

In general, it is best to structure the WBS on tangible, deliverable items, both software and hardware. Figure 6-2 shows two other WBSs, either of which could be used on the breath analysis project of Figure 5-4.

The more work packages you have in your project, the smaller and cheaper each work package becomes. However, the more work packages you have, the more money and time is spent in arranging for these to be properly interfaced with each other and managed. Conversely, if you have only one work package, there is no interfacing cost, but the task itself is large and expensive. Therefore, there is a happy midpoint that must be found by experience. In general, you should break your project into work packages sufficiently small so that each is understandable.

The WBS defines the work packages and will be tied to attendant schedules and budgets for the work performers. Thus, it is desirable for the lowest level packages to corres-

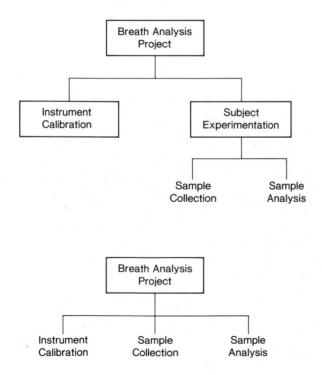

Figure 6-2. Two possible WBSs for project illustrated in Figure 5-4.

The WBS must be tied to time and money plans.

pond to small work increments and short time periods. It is often helpful to indicate who the task leader is by putting his or her name in the WBS box for the task. In any event, the WBS can clarify organizational responsibility on a project.

In addition to using the WBS for planning, use task authorization forms, such as in Figure 6-3, to clarify the statement of work for each task. If the task schedule or budget is a constraint, the task authorization form should have a blank for this. Note that the task authorization form has a block where the task leader accepts the task, which assures compliance with the Golden Rule discussed in Chapter 5.

TASK AUTHORIZATION		PAGE OF
TITLE		
PROJECT NO.	**TASK NO.**	**DATE ISSUED**
STATEMENT OF WORK:		
APPLICABLE DOCUMENTS:		
SCHEDULE START DATE:	COMPLETION DATE:	
COST:		
ORIGINATED BY: DATE:	ACCEPTED BY: DATE:	
APPROVED BY: DATE:	APPROVED BY: DATE:	
APPROVED BY: DATE:	APPROVED BY: DATE:	

Figure 6-3. Task authorization form.

Helpful Hints

In preparing a WBS, do not forget required tasks such as engineering analyses or trade-off studies that must be done but are not specific deliverable items. Also remember to include reports, reviews, and coordination activities. In fact, displaying them on a WBS is a good way to highlight that they are necessary and that resources must be devoted to them.

Fortunately, when a WBS is prepared, it tends to stress hardware integration activities. That is, junctions on the WBS frequently imply a hardware assembly or a test activity that must occur when these things are joined. Thus, a WBS again is useful for identifying an activity to which resources must be devoted.

There is also a slightly different way to create a WBS. It is a "costed WBS," and the dollar volume attached to each branch of the total project is also included in the WBS boxes (at least at the major headings). This kind of additional information, assuming large amounts of money are associated with large amounts of activity, will direct the manager's attention to those portions of the project that represent the most activity.

The costed WBS is often vital on an engineering project with rigid cost limits. In such a situation, it is often helpful to start with a costed WBS. A trial allocation of the rigid budget is then made to see how well each work package can be performed at that level of funding. Then adjustments are made to the funding allocations to provide the best overall balance of achievable performance within the budget limit.

If you can afford the time, it is desirable to have another person make a WBS for your engineering project, independent of yours, at least down to the third or fourth level. This will take only an hour or so and will highlight any discrepancies or oversights. This approach may suggest a more effective way to organize the required work. You will have to repay the favor on later projects, but that should help your organization by reducing problems on projects. In fact, some organizations require that two or more people independently prepare a WBS for a given engineering project before it can be approved.

After the initial WBS has been made, schedule planning can commence. The schedule planning may identify further items to add to the WBS. Although less likely, the same may occur as cost planning is done. The WBS is then revised to include these work packages so that everything on the WBS is finally tied to scheduled work packages and budgets and vice versa.

Others can help assure that your WBS is complete.

TYPICAL PROBLEMS

Vagueness in the SOW is a crucial problem in planning the performance dimension. For instance, the SOW may state that "appropriate tests will be performed." Who decides, and when, what is appropriate? The solution is to write a specific and detailed SOW.

Another problem is the blind copying of a prior project's WBS for a new project. When this occurs, we have not a WBS, but a waste of everybody's time—the people who prepare the WBS and the people who must read it. An engineering project's WBS should be prepared thoughtfully, not by rote, to increase the odds of project success.

A third problem occurs when you prepare a WBS for an engineering project and fail to revise it as subsequent schedule and cost planning reveal other work packages that are required.

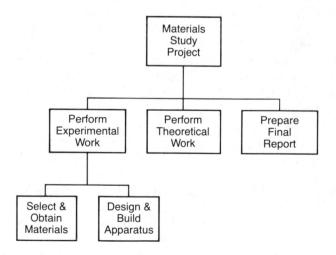

Figure 6-4.   WBS for materials study project.

MATERIALS STUDY PROJECT

The materials study project calls for experimental work, theoretical work, and a final report. To do the experimental work, you must test materials in an apparatus. The WBS might be as illustrated in Figure 6-4, and a task authorization for task B might be as illustrated in Figure 6-5.

TASK AUTHORIZATION	PAGE *1* OF *1*

TITLE *OBTAIN MATERIALS FOR MATERIALS STUDY PROJECT*

PROJECT NO. *83627*	TASK NO. *83627.02*	DATE ISSUED *3 JAN 84*

STATEMENT OF WORK:
OBTAIN the two materials To be tested in sufficient quantity to perform tests 3 times on each material.

APPLICABLE DOCUMENTS:
See our proposal (copy attached)

SCHEDULE

START DATE: *23 JAN 84*   COMPLETION DATE: *9 MAR 84*

COST: *$9,000.00*

ORIGINATED BY:	DATE:	ACCEPTED BY:	DATE:
APPROVED BY:	DATE:	APPROVED BY:	DATE:
APPROVED BY:	DATE:	APPROVED BY:	DATE:

Figure 6-5.   Task authorization form for the materials study project.

HIGHLIGHTS   *A work breakdown structure identifies all work packages required on a project.*

A co-worker's independently produced WBS for your project may identify omissions on your WBS.

The statement of work specifies what the customer will receive and when delivery will occur.

FURTHER READING

B. N. Abramson and R. D. Kennedy. *Managing Small Projects.* Redondo Beach, CA: TRW Systems Group, 1975.
> *Pages 12–17 of this breezy, short booklet give a graphic treatment of a WBS.*

R. D. Archibald. *Managing High-Technology Programs and Projects.* New York: Wiley-Interscience, 1976.
> *Chapter 7, section 5, is a good treatment of a WBS, which Archibald calls the project breakdown structure.*

J. A. Maciariello. *Program-Management Control Systems.* New York: Wiley-Interscience, 1978.
> *Chapters 4 and 7 provide a good treatment of the WBS, including the costed WBS.*

Scheduling Tools
and Time Estimating

This chapter is the first of two that deal with the second dimension of the Triple Constraint. The plan for the schedule dimension orders activities so you can identify the logical relationship between them. In general, there are three approaches to scheduling: bar charts, milestones, and network diagrams. I discuss each in this chapter and stress network diagram usage in the next chapter. This chapter also covers the important topic of time estimating, which must be done regardless of which scheduling tool is chosen.

BAR CHARTS

Bar charts, often called Gantt charts after H. L. Gantt, an industrial engineer who popularized them during World War I, are frequently used for scheduling. Figure 7-1 is a bar chart. The project is divided into five activities with a planned duration of twelve months. When the bar chart was constructed, five open bars were drawn to represent the planned time span for each activity. The figure also shows project status at the end of the sixth month. The shaded bars represent the forecasted span of the activities as of the end of the sixth month. Activity A was completed early. Activity B is forecast to be finished half a month late. Activity C is forecast to end approximately a month and a half early. The percen-

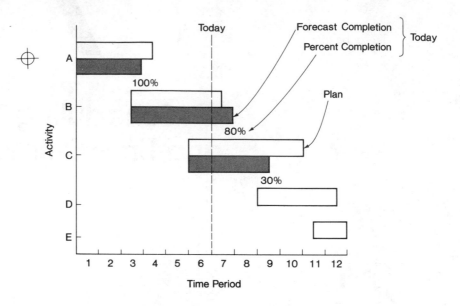

Figure 7-1.   Typical bar chart, illustrating a project with five activities at the six-month review.

Know the advantages and disadvantages of bar charts.

tage of completion for each activity in process is also illustrated. Activity A has been completed; B is 80 percent complete; and C is 30 percent complete.

Bar charts are simple to construct and easy to understand and change. They show graphically which activities are ahead of or behind schedule.

Offsetting these favorable features are some weaknesses, the most serious of which is that bar charts are essentially useless. Knowing the status of project activities gives no information at all about overall engineering project status because one activity's dependence on another and the entire project's dependence upon any particular activity are not apparent.

In addition, the notion of a percentage completion is difficult and is most commonly associated with the use of bar charts for measuring progress (which I discuss further in Part 4). Does the percentage completion refer to the performance dimension, the schedule dimension, or the cost dimension of the job? Unless an activity is linearly measurable, for instance, drilling a hundred holes in a steel plate, it is impossible to judge what percentage of it is complete. (Even in this simple case, the steel plate may have an internal defect, and the last drill hole might be through that defect, causing the plate to crack, at which point what was 99 percent complete now has to be done all over again.) Therefore, percentage completion becomes highly subjective or is frequently taken merely as the percentage of cost expended compared to total projected cost. In neither situation is percentage completion a useful number. Bar charts are much more useful as an indication of what has happened than as a planning tool to aid the project manager in making things happen properly in the future.

MILESTONES

A milestone schedule notes a few key events, called milestones, on a calendar bar chart. Milestones have been defined in various ways, but they probably are best defined as events clearly verifiable by other people or requiring approval before proceeding further. If milestones are so defined, projects will not have so many that the conclusion of each activity itself becomes a milestone.

A schedule that does not show task or activity interdependencies is useless by itself for planning.

The key to helpful use of milestones is selectivity. If you use only a few key events—perhaps one every three months or so—you will avoid turning milestones into pebbles (sometimes called inchstones) over which people are always stumbling. Some useful milestones might be, for instance, a major design review or a first article test.

When milestones have been defined, for instance, in the customer's request for a proposal or in your proposal document, listing them often helps in preparing your project plan. Having such milestones with attendant schedule and budget measures adds extra emphasis to a few key points of a project. But, in common with bar charts, milestone schedules do not clarify activity or task interdependencies. Thus, they must be used with other tools if they are used at all.

NETWORK DIAGRAMS

There are many forms of network diagrams, but the Program Evaluation and Review Technique (PERT) and the Critical Path Method (CPM) are the most common. "Network diagram" is a generic term for **PERT** and CPM diagrams, arrow diagrams, bubble diagrams, precedence networks, and many others. (It is not uncommon to have any network diagram

Networks indicate crucial interdependencies.

designated a PERT chart, whether or not it truly is.) Network diagrams are the recommended approaches to planning the schedule dimension for an engineering project; more detail on their use is provided in the next chapter. They identify the precedence conditions and the sequential constraints for each activity.

PERT and CPM emerged in different ways in the late 1950s. PERT is event oriented (that is, the event labels go in the nodes of the diagram) and has typically been used for aerospace and research and development (R and D) projects for which the time for each activity is uncertain. CPM is activity oriented (that is, the activity labels are placed on the arrows) and has been applied to the construction industry, in which there is typically a controllable time for each activity. There are now many hybrid forms of network diagrams that provide the best features of PERT and CPM. The network purist undoubtedly cares about the distinctions between these two, but in reality they are not very important. For simplicity in this chapter, I will assume PERT is always event oriented and CPM is always activity oriented.

Conventions

Arrows designate activities or tasks.

Figure 7-2 shows how an amplifier building project would be illustrated in PERT and CPM diagrams, and Figure 7-3 provides the symbolic conventions common to both. Activities are always shown as arrows with the start being the tail of the arrow and the completion being the barb. Events are shown as circles (or squares, ovals, or any other convenient closed figure). The event number is placed inside the closed

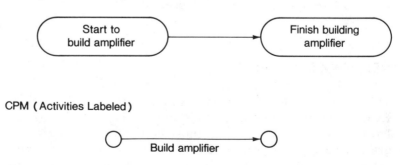

PERT (Events Labeled)

CPM (Activities Labeled)

Figure 7-2.   Examples of PERT and CPM.

Activities

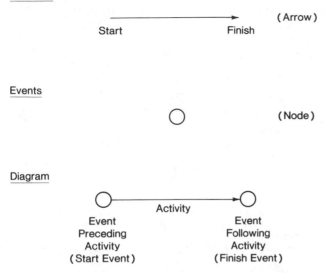

Figure 7-3. Basic network diagram conventions.

figure. Event numbers are required in computer programs, which are sometimes used to facilitate network information manipulation. In such cases, a higher number activity always follows a lower number activity. In computer programs, activities are not labeled by their name (that is, "build amplifier" is not called "build amplifier") but rather by the start and end numbers (that is, 5-10 is build amplifier, if 5 is the number of the start node and 10 is the number of the finish node). It is also conventional to place early and late date numbers or other information such as slack time within the nodes. Whenever node numbers are used, there should be a legend explaining which number is which. Using these conventions, a network diagram consists of a series of nodes and arrows connected to show the order of activities.

The upper drawing of Figure 7-4 depicts a schedule plan in which activity R must be complete before activity S can commence and in which activity T must be complete before activity U can commence. The middle drawing shows a schedule plan in which both activities R and T must be complete before either S or U can commence. The bottom drawing introduces the concept of a dummy activity, which is an activity requiring no work, that is, a precedence condition. It thus depicts a plan in which both activities R and T must be complete before activity S can commence and in which

A dummy activity is a precedence condition.

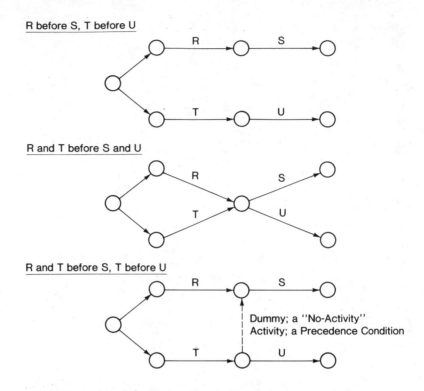

R before S, T before U

R and T before S and U

R and T before S, T before U

Dummy; a "No-Activity" Activity; a Precedence Condition

Figure 7-4.   Precedence requirements.

activity T must be complete before activity U can commence. Activity U does not depend on activity R because the dummy arrow points in the other direction.

Network Terms

Figure 7-5 illustrates three terms in network diagram usage. A burst node (node 20) is a node or an event at which two or more activities can be initiated after completion of a preceding activity. A merge node or event (node 50) is one in which two or more activities must be completed prior to initiation of the subsequent activity.

A dummy activity represents a dependency between two activities for which no work is specifically required. Dummies are also used to deal with an ambiguity that arises in computer-based network diagrams, also illustrated in Figure 7-5. As mentioned, in computer-based network diagram programs, the activity label is not the activity name but rather the number of the two nodes preceding and following it.

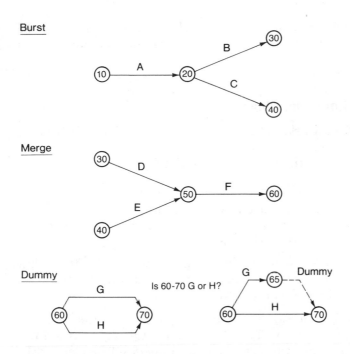

Figure 7-5.   Common network terms.

Thus, is activity 60-70 task G or task H in the lower left of Figure 7-5? Using a dummy task, one can make activity 60-65 task G and activity 60-70 task H, as shown in the lower right of Figure 7-5. There is thus a dummy activity, 65-70, also required to remove the previous ambiguity.

TIME ESTIMATING

Obviously, a schedule for an engineering project requires a knowledge (or estimate) of how long each activity or task will take. Because, by definition, the project has not been done before, there is a necessary inaccuracy in such time estimates. (The only way to guarantee meeting a time estimate is to make it infinitely long, in which case the project won't be authorized.) Thus, the goal is to try for reasonably accurate task time estimates, knowing that some will be short and some will be long. If some tasks on your project are identical or very similar to tasks previously performed, it is easier to estimate these. (As I said in the previous chapter, one of the goals of breaking the project into small work packages is to

Time estimates are usually inaccurate.

obtain understandable tasks, which also implies they can be estimated.) There are two time estimating techniques you may wish to use: PERT and pragmatic.

PERT Time Estimating

PERT networks originated in projects characterized by uncertain times for activities. This problem was dealt with by requiring three time estimates for each activity:

1. The most probable activity time (T_m)

2. The optimistic activity time, namely, the shortest time that might be achieved 1 percent of the time such an activity was carried out (T_o)

3. The pessimistic activity time, namely, the time that would be exceeded only 1 percent of the time such an activity was carried out (T_p)

As Figure 7-6 shows, this permits calculation of the expected time for the activity (T_e). The basis for this calculation, unproven and unprovable, is the rule applied in PERT networks and is, in fact, a reasonable way to estimate. You can also

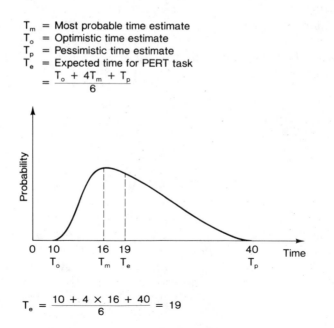

T_m = Most probable time estimate
T_o = Optimistic time estimate
T_p = Pessimistic time estimate
T_e = Expected time for PERT task
$$= \frac{T_o + 4T_m + T_p}{6}$$

$$T_e = \frac{10 + 4 \times 16 + 40}{6} = 19$$

Figure 7-6. PERT time estimating.

calculate the uncertainty of that time, which is called the standard deviation (σ). The calculation is illustrated in Figure 7-7.

$$T_o = 2$$
$$T_m = 5$$
$$T_p = 14$$

$$T_e = \frac{(1 \times 2) + (4 \times 5) + (1 \times 14)}{6} = 6$$

$$\sigma = \frac{1}{6}(T_p - T_o) = 2$$

Figure 7-7. PERT time uncertainty (σ-standard deviation) for a single event.

Figure 7-8 shows how to figure the expected time for a path and standard deviation of the path's expected time. The significance of the calculated standard deviation is the same as with the normal (Gaussian) probability distribution: Two-thirds of the time, the work will be completed within plus or minus one standard deviation of the expected time; 95 percent of the time, it will be completed within two standard deviations; and 99 percent of the time, it will be completed within three standard deviations. This kind of calculation can be important and helpful if there will be a cost penalty for lateness because you can estimate the likelihood of being late.

PERT time estimating is useful when time schedule is critical.

Figure 7-9 illustrates the three time estimates for a particular path containing three activities. Completion of the calculations for this case would show that the expected time is twenty-nine days and the standard deviation is six days.

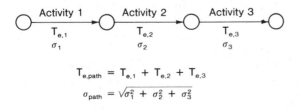

$$T_{e,path} = T_{e,1} + T_{e,2} + T_{e,3}$$
$$\sigma_{path} = \sqrt{\sigma_1^2 + \sigma_2^2 + \sigma_3^2}$$

Figure 7-8. PERT expected time and uncertainty for a path.

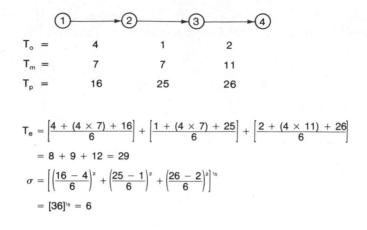

$$T_e = \left[\frac{4 + (4 \times 7) + 16}{6}\right] + \left[\frac{1 + (4 \times 7) + 25}{6}\right] + \left[\frac{2 + (4 \times 11) + 26}{6}\right]$$

$$= 8 + 9 + 12 = 29$$

$$\sigma = \left[\left(\frac{16 - 4}{6}\right)^2 + \left(\frac{25 - 1}{6}\right)^2 + \left(\frac{26 - 2}{6}\right)^2\right]^{1/2}$$

$$= [36]^{1/2} = 6$$

Figure 7-9. Path with three time estimates for each activity.

Thus, the project's completion would be between the twenty-third and thirty-fifth day two-thirds of the time.

Because it requires a lot of effort to make three time estimates and even more time to calculate the expected time and standard deviations, you would normally do this only with a PERT network that was being put on a computer. (And, even with a computer available, you might prefer to use pragmatic time estimating.) Nevertheless, it is important to appreciate the technique and apply it where warranted.

Pragmatic Time Estimating

I recommend pooled pragmatic judgment for time esti-mating. The task leader, project manager, and one to three others should discuss the task and arrive at a judgment as to what the schedule should be. The task leader is there because of the Golden Rule. The project manager is there to provide balance with other project time estimates. The others are there to bring expertise and experience to bear.

As a practical matter, the project manager, task leader, and one to three others cannot hold discussions on every task on a large project because there simply is not enough time. In such a large project, this is the goal at which to aim, and a practical compromise is for the project manager to have several deputies to represent him or her in these task estimat-ing meetings.

The goal of such a group estimating meeting is to arrive at a sensible consensus for how long the task will take when it is run in the intended way. If the task leader is a junior engineer, he or she will not normally be able to complete the task as quickly as a senior engineer (who may be one of the consulted experts). Sometimes the reverse is true—a junior mechanical engineer may be able to complete a design very quickly using computer-aided design tools, which older engineers have never learned to use.

If a task, as distinct from the engineering project of which it is a part, is identical (or very similar) to a previously completed task, then the experience on the prior task is a good estimating guide. But be certain that there are no meaningful differences that invalidate the relevancy of previous experience.

A logical sequence for estimating a new task is to (1) determine how many days the previous, similar task required and how many personnel worked on it by consulting existing company project records, (2) decide how much more complex the present project is to arrive at a time duration and personnel multiplier, and (3) determine the cost of the new task by multiplying the person days by the appropriate labor rates. This assumes such records exist and underscores the importance of retaining project histories. If no such records are retained, then memory is all that can be used, and memory may differ from one person to another.

Base the time estimate for a task on who will do the work and how they will do it.

TYPICAL PROBLEMS

In many ways, the worst schedule dimension planning problem is to avoid the indicated scheduling problems. For instance, a completed network diagram may show that required materials will not arrive early enough. This conflict is often avoided or dismissed by saying this can be adjusted later. Maybe it can, but that is hoping for luck to save your project schedule. The solution is to admit the problem exists and revise the schedule to overcome it—now, not when there is no longer time to correct the problem and maintain your schedule.

Sometimes the project manager or higher management does not like the overall estimated length of the project and wants it reduced. Such a reduction is a

continued

problem if time is merely cut out of a task without changing the task's work plan to reflect how this reduction can actually be accomplished.

It is difficult to obtain accurate time estimates for things not done before. As suggested earlier, getting a few people together, including especially those who will be responsible for the activity, and pooling judgments is the best solution to this problem.

MATERIALS STUDY PROJECT

For the materials study project, in terms of the schedule, assume the following:

- Experimental task
  1. Materials
     A. Select (three weeks)
     B. Obtain (six weeks)
  2. Apparatus
     C. Build (twelve weeks)
     D. Debug (two weeks)
     E. Use to conduct the experiment (eleven weeks)
     F. Document the design (six weeks)
- Theoretical task
     G. Review the literature (six weeks)
     H. Theoretical study (ten weeks)
     J. Report of theoretical results (five weeks)
- Final report task
     K. Final report (one week)

The bar chart for this is Figure 7-10.

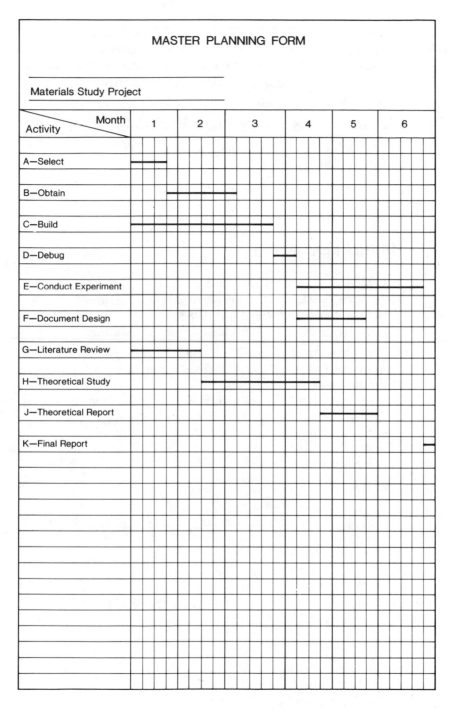

Figure 7-10.   Bar chart for materials study project.

HIGHLIGHTS

Although easy to make and understand, bar charts and milestones alone are inadequate for schedule planning because they do not show how one activity depends on another.

Network diagrams show activity interdependencies.

The most common network diagram forms are PERT and CPM.

Time estimating is necessarily inaccurate and can best be done pragmatically.

FURTHER READING

J. J. Moder, C. R. Phillips, and E. W. Davis. *Project Management with CPM, PERT and Precedence Diagramming.* New York: Van Nostrand Reinhold, 1983.

This recent book is devoted to project scheduling.

J. Mulvaney. *Analysis Bar Charting—A Simplified Critical Path Analysis Technique,* U.S. ed. Washington, D.C.: Management Planning Systems, 1977.

This is a simple, concise treatment of precedence diagrams, which are easy to use as scheduling tools on small projects.

PERT and CPM Network Diagrams

This chapter continues the previous chapter's discussion of scheduling techniques. First I clarify why a network diagram, especially a time-based CPM diagram, is the most useful tool for planning an engineering project's schedule. Next I provide some helpful hints for constructing a network diagram. Then I discuss how a time-based CPM diagram can clarify the time when individual tasks can or must start, which is crucial for coordinating work. I also discuss some alternative ways to portray network diagrams.

WHY USE A NETWORK DIAGRAM?

An Illustrative Situation

Consider the engineering project situation illustrated by the bar chart in Figure 8-1. You are the project manager for this project with eight activities or tasks. At the end of four months, you are conducting a project review (denoted by the triangle). Task managers provide status reports (denoted by the shading) showing that tasks B and C are two months late, A and D are one month late, and E is on schedule. The impact of these delays on the entire project's completion is not clear. (For simplicity in this example, assume that task status is precisely measurable—for instance, by counting the number

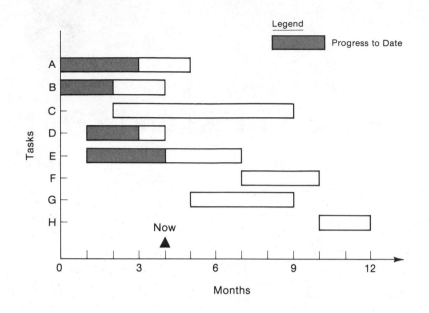

Figure 8-1.   Bar chart showing tasks A, B, C, and D to be late at four-month project review.

of holes drilled or drawings completed.) Your chief concern at this point is whether the entire project is late.

First you might break the tasks into their subtasks or subactivities, as in Figure 8-2. This provides additional information but still does not tell us whether the project will be late.

A PERT Network Diagram

Networks have more information than bar charts.

At this point, a network diagram for the project can be examined (Figure 8-3). This contains more information than the bar charts because it shows the interrelationship (precedence) between different tasks. I have labeled each event in its node with the completion of the designated subtask. Thus, at the top of the diagram, you can see that activity D must be completed before activity G.

Figure 8-3 also shows the problem with a PERT network: The activities per se are not illustrated; that is, there is no arrow uniquely associated with activities G, H, or C_3. This will always be the case where two or more arrows come to a single node, that is, at all merge nodes. This is not a problem for a skilled PERT practitioner, but it does seem to present an unnecessary conceptual difficulty. When the activities are not explicitly shown on a diagram, it is more difficult for the

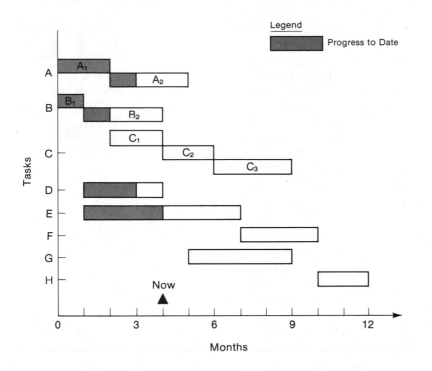

Figure 8-2. Bar chart (from Figure 8-1) with subtask breakdown.

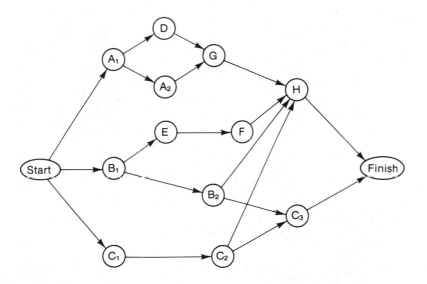

Figure 8-3. PERT network diagram for project of Figure 8-1. Each node designates the completion of the indicated task or subtask.

project manager and others to visualize them and their relationship.

A project manager's ability to influence the course of his or her project depends on his or her ability to influence the work on a given task or activity. One of the few things a project manager can do is change the allocation of resources devoted to a particular activity. Thus, the lack of each activity's explicit visibility in a PERT diagram may be troubling.

A CPM Network Diagram

This PERT problem can be eliminated if we go to a CPM network (Figure 8-4), which shows all activities by labels on the arrows. It clearly indicates the precedences. The requirement of a dummy activity, a "no-activity" activity, is to indicate that the completion of activities B_2 and C_2 (as well as activities F and G) must precede the start of activity H. But in this CPM representation, the merge nodes are not single activity completion events. For instance, the node to which D and A_2 arrows come would now have to be designated the completion of both D and A_2. Frankly, this CPM duality does not trouble me as much as PERT's lack of activity emphasis and clarity. In fact, the duality may actually be helpful by emphasizing that both D and A_2 must be completed before G can commence.

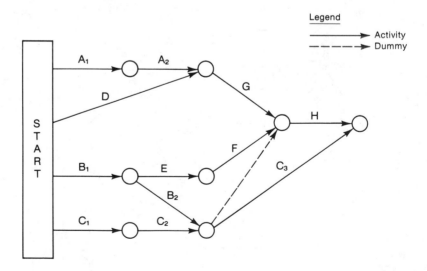

Figure 8-4. CPM diagram with each subtask activity and one precedence condition (or dummy).

Figure 8-5 illustrates the next step in using the CPM diagram: redrawing it to a time scale in which the horizontal projection of each arrow is proportional to the amount of time required for its activity. Doing this reveals that one path (B_1, E, F, H) is longer than any other. This is called the critical path. It may also be identified as the path that contains no slack time (amount of time available on a path that is the difference between that required on the critical path and that required on the particular activity path with slack time).

The critical path, which indicates the shortest time in which the project can be completed, has no slack.

Figure 8-5 is drawn with each activity shown starting at the time it was scheduled in Figure 8-2. Noting the task progress status and current date, Figure 8-6 begins to provide direct information as to the implications of the delay on activities A, B, C, and D. Figure 8-7 is a redrawn version of Figure 8-6 in which the project manager has taken advantage of the slack time. That is, all late (delayed) activities are drawn to show the work remaining to be done, and subsequent tasks are thus rescheduled in several cases.

Thus, although activities A, B, C, and D are in fact later than planned, the project has not yet suffered any irretrievable schedule slippage. But the project now has two critical

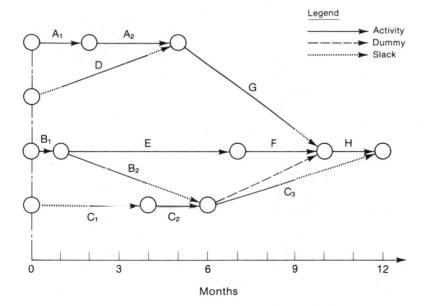

Figure 8-5. Time-based CPM diagram, drawn on assumption of task schedules shown in Figure 8-2. (Note that the start node representation is an alternative to that used in Figure 8-4.)

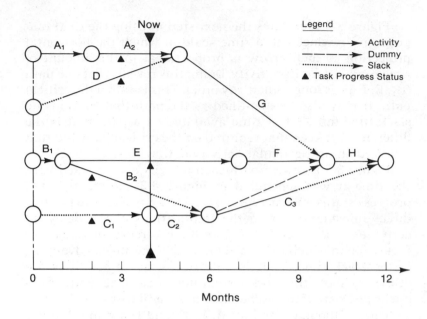

Figure 8-6. Figure 8-5 with task progress status from Figure 8-2 denoted by solid triangles.

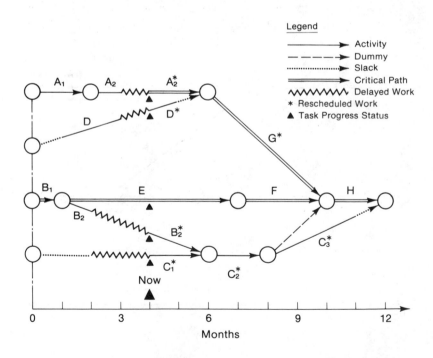

Figure 8-7. Time-based CPM network, redrawn to show delayed work and rescheduled activities, permitting project to be completed in accordance with original twelve-month schedule.

paths, whereas it previously had only one. That is, there is no longer any slack in the upper branch (A_2 and G). Because there is still one month of slack on the lower path (task C), perhaps some of the resources allocated to it (or to task B_2) might be redeployed to one of the other critical path activities. It is vastly more difficult to complete a project on time with more than one critical path, and it is unlikely that this project will be completed on schedule, although it is not yet irretrievably lost.

Another Example

Figures 8-8 and 8-9 are a CPM diagram and two versions of bar charts for a house-painting project. The network diagram clearly contains far more information than either of the bar charts—for instance, the dependency of D_1 and D_2 on A_3.

HELPFUL HINTS

One frequently asked question is, "How do I start a network diagram?" One answer is, "With lots of scrap paper." But the best way to start is with the work breakdown structure. From the WBS, you can start the network diagram from either the

Include every element in the WBS in the network diagram.

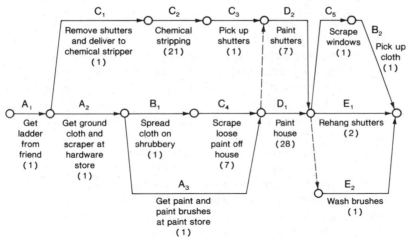

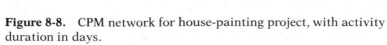

Figure 8-8.   CPM network for house-painting project, with activity duration in days.

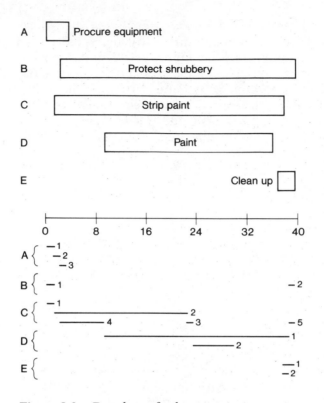

Figure 8-9. Bar charts for house-painting project, again revealing lack of task interdependency information.

beginning or the end of the project. There are frequently somewhat obvious large subnets you can quickly put down on a piece of scrap paper. As a general rule, it is probably best to start from each end with scrap paper and sort out the connectedness in the diagram where there are activities in progress simultaneously. As I said in Chapter 6, you may now find some work tasks that should be inserted into the WBS. If so, revise the WBS accordingly.

You can then transfer the entire diagram to a clean piece of paper. It is probably helpful to do this with a time base and with the presumption that each activity starts at the earliest possible time. For this initial version, I recommend that you get the people who will be responsible for each activity to estimate how long it will take to carry it out on a normal work basis. When you put these time estimates onto the network diagram, it may become apparent that the entire project will take too long. At this point, you can identify particular activities that may be candidates for time compression, that is, tasks you believe can be done faster.

An alternative to time compression is the parallel scheduling of activities, for instance, software and hardware development. There may be increased risk in such a schedule, but that may be the lesser evil. The final step in creating a time-based CPM diagram is to estimate the time of each activity as accurately as possible, using the techniques described in the previous chapter.

As shown in Figure 8-10, network diagrams may require crossovers of lines. This is to be expected. Although some diagram rearrangement may get rid of crossovers, it may also distort the logical relationship of groups of activities, for instance, all those being carried out by one department being within a general band of the diagram. If activities A, B, C, D, E, F, and G in Figure 8-10 are performed by one section, the upper diagram, which has two crossover intersections, would be preferable to the lower one. Thus, there are cases in which increased use of crossovers will be clearer.

The project manager should construct a network diagram of perhaps three dozen activities or up to five dozen if required. Such a diagram can normally be drawn in less than two hours and will fit onto a standard seventeen-by-twenty-two-inch sheet of paper. If some activity in this network is

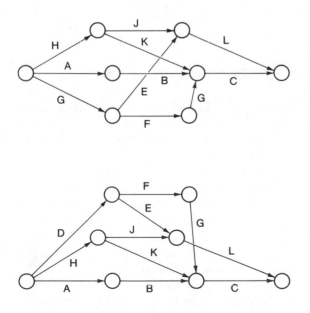

Figure 8-10.   CPM network diagram with crossovers.

If you can't make a network diagram, you can't run the project.

very large, its activity manager can make a network diagram for it. In this way, with a few hand-drawn networks of a few dozen activities each, large projects can be handled without the use of a computer-based network system.

Some project (and other) managers resist the use of network diagrams because they consider them complex or because they lack a computer-based project planning network program. This is a serious mistake. It is not the network diagram that is complex; it is the project itself that is complex. In fact, if you can't draw a network diagram for your project, that should be a clear danger signal that you do not understand your project.

I am not opposed to the use of a computer-based planning network program to assist with the mechanics of network usage. In fact, computer programs have great value in determining resource requirements quickly. But a computer-based network diagram program is not required to manage most projects, and the lack of such computer assistance is no excuse for not using a network diagram.

Always use a network diagram to plan the schedule dimension, even if you do not show it to your management or customer.

Very large projects, however, will normally require a computer-based system, and many are available. But you usually lose the "hands on" feel you can get by drawing your own network diagram. In general, avoid the computer until you are forced to use it. The time required to redraw a network periodically to keep it current may be less than what you would spend correcting incorrect data entry into the computer.

EARLIEST AND LATEST START AND FINISH TIMES

To introduce this important benefit of network diagrams, consider Figure 8-11. This is a PERT version, with emphasis on the nodes. It is conventional to assume that projects start at time equal to zero. The earliest time (T_E) you can emerge from the start node is zero. On the critical path, adding the activity time (in this case, 10), the earliest time you can get to the finish node is 10. Latest and earliest are always the same on nodes for the critical path; hence, $T_L = 10$ at the finish node and $T^E = 0$ at the start node. Off the critical path, the earliest you can reach the upper node is the earliest you can leave the start node plus the activity time on that path (in this case, 2).

Latest times are calculated by working backward. Thus, T_L at the finish node (10) less the activity time (6) determines that the latest time you can leave the upper node (without

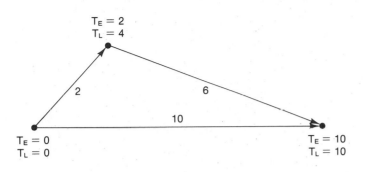

$T_E = 2$
$T_L = 4$

2

6

10

$T_E = 0$
$T_L = 0$

$T_E = 10$
$T_L = 10$

Figure 8-11. Earliest and latest start.

Source: "Project Management." Lecture notes by Milton D.
Rosenau, Jr. Copyright © 1981 by the Association for Media-based
Continuing Education for Engineers, Inc. (AMCEE). Reprinted by
permission.

delaying completion) is $T_L = 4$. The difference between $T_F = 2$
and $T_L = 4$ at this upper node (2) is the slack, or float, on the
upper path.

Now consider a CPM activity-oriented network diagram,
which I prefer. The entire project is always assumed to start
at time zero. Thus, the start of each activity that emerges ·
from the start node has zero as its earliest start time (E_S). The
earliest finish time (E_F) for each of these initial activities is
the duration of the activity itself (Figure 8-12A). Earliest start
and finish times are calculated by proceeding from the start
node to the finish node. In Figure 8-12A, activity duration is
shown by the number above the middle of the activity
arrows. The earliest finish of an activity is equal to the activity duration plus the earliest start. At the merge node, the
earliest start of the following activity is the higher of the
earliest finishes for the preceding activities. On the critical
path, the earliest finish at the finish node is both the minimum project duration and the latest finish for that activity.

Figure 8-12B shows how to calculate latest finish (L_F) and
latest start (L_S) times for each activity. Calculation com-
mences at the finish node and proceeds backward to the start
node. In Figure 8-12B, on the critical path, the latest and
earliest times are equal. The latest start of an activity equals
the latest finish of that activity minus the activity duration.
At the burst nodes, the latest finish of the preceding activity is
the lower of the latest starts for the following activities.

Table 8-1 shows the kinds of data provided by a typical
computer printout for the project illustrated in Figure 8-12.
Although less graphic, these data reveal the same informa-

*The difference
between earliest
and latest times
at a node
indicates the
amount of slack.*

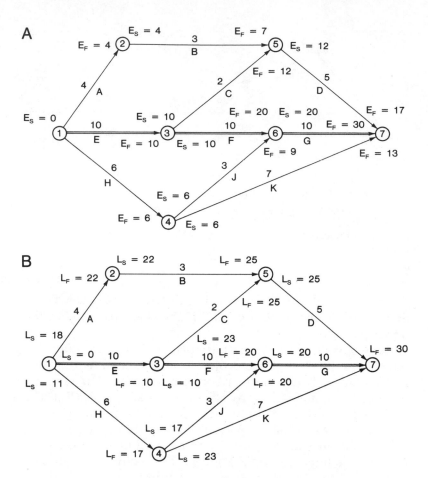

Figure 8-12(A). Earliest Start (E_S) and earliest finish (E_F) calculation. **(B).** Latest start (L_S) and latest finish (L_F) calculation.

tion. Normally, all the earliest and latest information would be on one diagram. This is illustrated in Figure 8-13. Note that the use of vertical dashed lines, without any dependency arrow indication, permits a node to be drawn in more than one location. This permits spatial separation of activities, thus providing additional open space on the network diagram.

BAR CHART FORMATS OF CPM DIAGRAMS

Although it is often said that CPM diagrams are difficult to use during project reviews and management briefings because of their apparent complexity, many organizations insist

TABLE 8-1. Typical Data Provided in a Computer Printout for Computer-Based Network Reporting. (Use with Figure 8-12.)

| Event | | | | Start | | Finish | | |
Start	Finish	Description	Duration	E	L	E	L	Slack
1	2	Activity A	4	0	18	4	22	18
1	3	Activity E	10	0	0	10	10	0
1	4	Activity H	6	0	11	6	17	11
2	5	Activity B	3	4	22	7	25	18
3	5	Activity C	2	10	23	12	25	13
3	6	Activity F	10	10	10	20	20	0
4	6	Activity J	3	6	17	9	20	11
4	7	Activity K	7	6	23	13	30	17
5	7	Activity D	5	12	25	17	30	13
6	7	Activity G	10	20	20	30	30	0

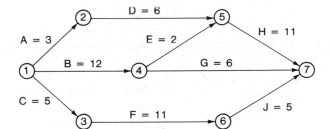

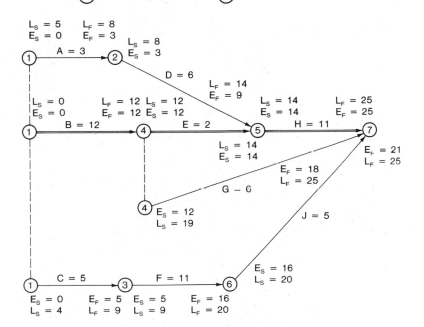

Figure 8-13. CPM earliest and latest start and finish.

they be used for these purposes. There are at least two ways to make management personnel attending such reviews comfortable with CPM presentations. First, activities can be displayed in a bar chart, indicating their planned time, the earliest start and latest finish, and slack, as shown in Figure 8-14. A second approach is to use vertical connections between activities that are dependent on one another to illustrate that dependency (Figure 8-15). However, to use the bar chart representation of a CPM diagram, you must start with the CPM diagram, not with a bar chart. Figure 8-15 is really a time-based CPM diagram with all the activity arrows drawn horizontally.

TYPICAL PROBLEMS

There is an interaction of schedule with resources applied. A good designer is usually fast and accurate, and the later fabrication activities may thus be fairly quick. Conversely, a poor or junior draftsperson may be slower and less accurate, and subsequent activities may take longer. The plan reflects what you intend to
continued

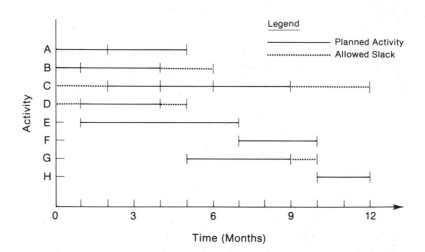

Figure 8-14.   Bar chart representation of network diagram from Figure 8-5.

do, but is not necessarily what you will do. Another aspect of this resource interaction is that having two people on one task is not necessarily as productive as having one person taking twice as long because the two people must spend time communicating with each other.

MATERIALS STUDY PROJECT

This example project can be shown in a network diagram (Figure 8-16). Figure 8-17 is a bar chart representation of the CPM network diagram for this project.

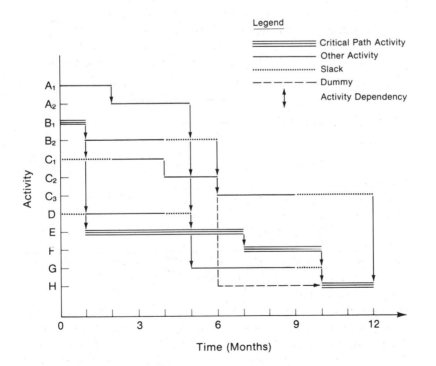

Figure 8-15. Bar chart representation of network diagram from Figure 8-5, with task dependency illustrated.

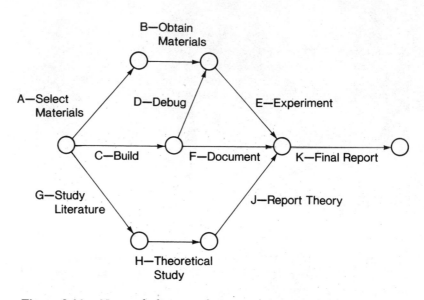

Figure 8-16. Network diagram for materials study project.

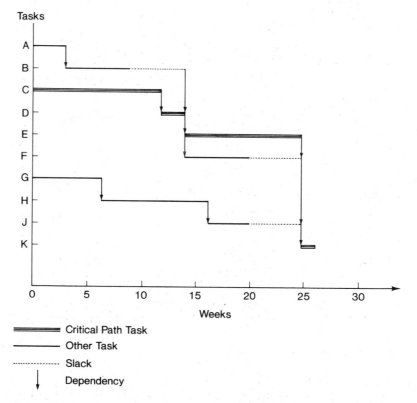

Figure 8-17. Bar chart representation of CPM diagram for materials study project.

Always use a network diagram, which must include every element in the WBS, to plan the schedule dimension of the Triple Constraint.

The most useful network diagram is a time-based CPM diagram because it explicitly shows each activity. In some cases, it may be helpful to present the time-based CPM diagram in a bar chart format.

A network diagram can provide earliest and latest start and finish times for each project activity.

FURTHER READING

J. Gido. *An Introduction to Project Planning.* Schenectady, NY: General Electric, 1974.
    This is a simple, concise treatment of network diagrams.

D. W. Lang. *Critical Path Analysis,* 2nd ed. London: Hodder and Stoughton Paperbacks, 1977.
    This is a good book on networks.

J. D. Wiest and F. K. Levy. *A Management Guide to PERT/ CPM,* 2nd ed. Englewood Cliffs, NJ: Prentice-Hall, 1977.
    This is a clear and complete book devoted to PERT and CPM.

Planning the Budget

Having a cost plan can help you avoid a situation where actual project cost overruns the estimate or you fail to get the job because you overestimated costs during the proposal and negotiation phase. This chapter tells you how to estimate costs and describes several of the most commonly used cost systems.

COST ESTIMATING

Costs may be stated only in terms of the number of labor hours required, a situation not uncommon in a research group in which a certain number of labor hours have been allocated to a particular project. Cost is more commonly stated in dollars (or yen or marks), however, which entails converting labor hours into dollars. Different hourly rates typically prevail for different seniority levels, and the cost of nonlabor elements (purchases or travel, for instance) is also included. Figure 9-1 shows one way to summarize and total time-phased labor and nonlabor estimates for a task. This figure illustrates the main elements of any successful cost estimating system: Estimate *labor hours* (perhaps by category) and *nonlabor dollars* for each task in each involved department or group.

Cost is, of course, necessary for planning an engineering project, both to sell and manage the job. In general, do not plan costs in detail greater than what you will receive in

PROJECT _MATERIALS STUDY_____ TASK _B-OBTAIN MATERIALS_ DEPARTMENT _MECHANICAL_

COST ELEMENT			HOURS EACH MONTH						TOTALS	
			1	2	3	4	5	6	HOURS	DOLLARS
LABOR	SR. PROGR.-ANAL.	$25/hr.	8	4	2				14	350
	JR. PROGR.	$20/hr.		40					40	800
	QA ANAL.	$15/hr.								
	CLERK	$10/hr.								
			DOLLARS EACH MONTH							
DOLLARS	LABOR COST									1150
	OVERHEAD	100%								1150
	DIRECT NONLABOR		200							200
	PRIME COSTS									2500
	G & A	15%								375
	TOTAL COSTS									2875
	PROFIT	20%								
	TOTAL BILLING									

ASSUMPTIONS _____

PREPARED BY _____ DATE _____ APPROVED BY _____

Figure 9-1. Typical task cost estimate.

accounting cost reports. There is no point making cost plans out on a daily basis if the organization's cost reports are furnished biweekly or monthly. Cost plans, regardless of how they are arrived at, should typically be summarized in monthly periods corresponding to expense reporting. In counting such things as travel cost or computing hours, however, work with hours or days of travel in estimating and sort these into monthly periods.

Plan costs to the level of detail to which they will be reported to you.

Just as with the schedule dimension plans, there are inaccuracies inherent in cost estimates, and these must be expected and tolerated. But tolerating such inaccuracies does not mean encouraging them. The goal is to be as accurate as possible and to recognize that perfection is impossible.

Techniques

"Forecasting" and "estimating" are frequently used interchangeably to refer to preparing a plan for the cost dimension. Actually, the dictionary definitions of these words are somewhat different. In engineering project management, we are talking about the amount of money (or time) expected to be required to complete a piece of work.

If you were asked to estimate *pi*, you could do so as accurately as you wish because *pi* is a known quantity (3.14159 . . .). If, however, you were asked how long an untelevised football game will take, you would probably reply two and a half hours. You are now estimating a future event's duration based on similar previous events. You might have made this estimate by looking up the time for the longest and shortest football games ever played and by noting the times of all other football games for which durations were recorded. You would have learned that the vast majority of football games took between two and a quarter and two and three-quarter hours; therefore, two and a half hours is a reasonable estimate.

Actually, the football game you will see will not take two and a half hours. The probability of your estimate being correct is essentially nil. The only way to *guarantee* that actual costs do not exceed your estimate is to make the cost estimate very, very high—which will probably mean there is little likelihood of getting the project authorized. The fact that you are not going to be right means you should become accustomed to being wrong and should not be afraid of it. But it does not imply that you should not try to be accurate. Despite these hazards, the goal in estimating is to have a meaningful plan for your project, one you can use to sell the project proposal to your customer, explain your actions to your boss, and provide enough resources to do the job successfully.

Schedule first, estimate second.

There is no point in attempting to estimate a budget for an activity until you have established its duration. In addition, you should understand the preceding and following activities in order to define better the activity you are estimating. Such understanding may clarify that a following activity is farther downstream than it first appears. If so, the activity you are estimating probably is longer, and therefore costs more, than you first thought.

Estimate the cost of each task.

You do the estimating by breaking the project into tasks and activities, using the WBS and network diagrams. The budget of any large activity is the sum of the smaller tasks that compose it, as shown in Figure 9-2. In general, use as much detail as possible. Every task in the WBS should probably have an individual task estimate (such as Figure 9-1) prepared by the responsible task manager.

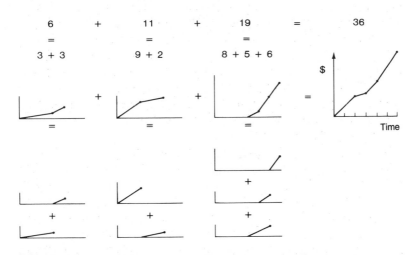

Figure 9-2. A project's cost is the summation of costs for all project tasks.

Shortcuts

There is a variety of means to prepare cost estimates. Using as much detail as possible is commonly called the "bottom up" method. The major project is divided into work packages small enough to allow accurate estimation. The project estimate is the sum of the estimates for all the individual work packages.

There are shortcuts to estimating some of the small work packages. You can use similarities to and differences from other tasks to shortcut a complete level of detail for a second task. Or you can use ratios or standards to relate one small task to another.

Whenever you use the bottom up technique, judge it against a "top down" estimate. The top down estimate is done first, quickly and judgmentally, and then it is set aside. For instance, assume the bottom up estimate comes to $10 million. Your top down estimate, which you now retrieve, is $5 million. Go back and look at each individual work package in the bottom up estimate to find out where the excess costs arise. Examine each package to discover to what extent there has been an incorrect assumption as to the amount of work called for. Or your top down estimate may indicate that the total job should cost $20 million. Explore the details to find out what has been overlooked or what unwarranted simplifying assumptions were made. The role of the top down esti-

To make a project cost estimate, add detailed task estimates for each work department and adjust them if the overall summation seems unreasonable.

mate, which is obviously not accurate, is to provide a point of view from which to scrutinize the bottom up estimate.

Parametric Cost Estimating

To introduce this technique, consider the situation in which you need to obtain some software to support your engineering project.

Figure 9-3 illustrates a relationship between project cost and a computer program's lines of code that we might discover if we examined a large number of previous computer software projects. The shaded area in the figure would surround a cloud of points, each of which represents a particular project outcome.

If we examine these projects in more detail, we could ask how many separate modules were involved. Then the data points could be separated, and we might find a trend as in Figure 9-4, where the lines are the centroids of smaller clouds of data points.

This is the underlying notion of parametric cost estimating. Historical data for many factors are used:

Project magnitude—How big is it?

Program application—What is its character?

Level of new design—How much new work is needed?

Figure 9-3.   Relationship between project cost and lines of program code for a large number of computer software projects.

Resources—Who will do the work?

Utilization—What are the hardware constraints?

Customer specifications and reliability requirements— Where and how are these used?

Development environment—What complicating factors exist?

Then, if your project falls within the range of these historical data, you merely insert your project's parameters into the multidimensional regression model to estimate your project's cost.

There are many such parametric cost estimating models. RCA's PRICE H, primarily applicable to government hardware projects, is accessible for interactive on-line use. The estimate you get from using such a parametric model is only an estimate and will be inaccurate if your input data (for instance, the resources) are poor estimates.

Parametric model estimates are only as accurate as the input data used.

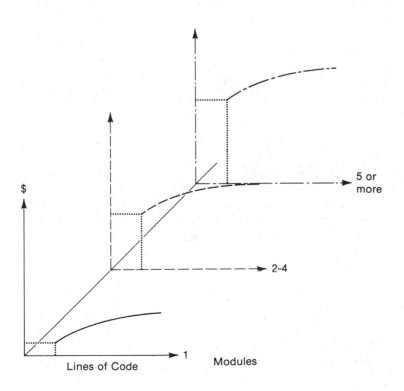

Figure 9-4. Trends of Figure 9-3 data when the number of program modules is considered.

Cautions

Cost estimates are frequently made prematurely, before the work performance and time schedule are fully understood and defined. Such cost estimates must then be redone or adjusted when the performance and schedule are set. It is clearly more efficient to do the cost estimating after the other two dimensions are defined.

There is a danger signal to which project managers must be alert when receiving cost estimates: the person-month dimension to describe the labor requirement. This is a danger signal because it frequently indicates a snap judgment on the part of the estimator (and not because it is an inappropriate measure, although hours seem preferable). The person who can do the task in two months may not be available when he or she is needed. Or a task that can be done by one person in three months (three person-months) may require four people if it must be completed in one month (four person-months).

On lengthy projects, have an inflation hedge.

A pertinent issue today is how to plan for inflation. Such planning can be done only with great difficulty and caution. Unfortunately, there are no guarantees in dealing with the future; so some method of coping must be adopted. Make the best estimate of labor hours, regardless of when the activity will occur. Then estimate the rates for these labor hours and the dollar amounts for nonlabor ingredients in current dollars. You can then apply inflation factors for future years to these numbers by consulting with your organization's financial planners.

PROJECT COST SYSTEM

To use project cost reports, you must have a project cost accounting system, which is a means to accumulate costs by project and project activity or task detail. The elements of such a system are:

- Labor
  People in your own department
  Other people in company
- Overhead burden
- Nonlabor
  Purchases
  Subcontracts
  Travel
  Computer charges
- General and administrative burden

There are many variations in detail for engineering project cost systems, some of which are shown in Table 9-1. I am making the simplifying assumption that the organization has three projects (A, B, and C), which all start at the beginning and will end at the close of the fiscal year. It does not matter whether the work is for the company or an outside client or whether it is to be paid for by the sale of goods or contract billing. The point is that there has to be some way of allocating the cost of these three projects to different customers or product lines. The table illustrates four methods of allocating these costs to the three projects.

In method 1, the direct labor and direct nonlabor are allocated to the project, and these are summed to provide a direct total. All the burden and overhead accounts are then lumped and apportioned to each project in proportion to the direct total expenses. In method 1, these are equal, and the billings to each of the three projects would be equal.

In method 2, the direct labor and direct nonlabor are treated as before, but the overhead portion is allocated to each project in accordance with the amount of direct labor it requires. Nevertheless, the general and administrative (G and A) expenses are allocated to the projects in accordance with the direct total, as in method 1. In this case, the billings to the projects are not equal. Project A is more than B, which is more than C.

In method 3, the overhead is treated as in method 2, and the direct nonlabor is treated as in both methods 1 and 2. But in this case, all these items and direct labor are joined to come up with a prime cost, and the G and A burden is allocated in proportion to that. In this case, we arrive at a still higher amount of billing for project A.

Method 3 and its variants are the most common project cost systems.

Method 4 is one of the many common variations of method 3. Direct labor and overhead are treated as in method 3, but purchases are subject to a material handling charge. (In method 3, this material handling charge is included in G and A; in method 4, it is pulled out of G and A and assigned to the projects in proportion to their required purchases.) But other direct nonlabor, in this case, travel, is not allocated a handling fee, as illustrated in project C. This results in a still different prime cost. Finally, the G and A expenses are again distributed, and a still different billing arrangement is arrived at.

All these methods, and others as well, are used. The project manager must understand his or her company's method in order to know when to use subcontract help and when to use in-house, direct labor.

It is also important to understand any subcontractor's

TABLE 9-1. Four Common Project Cost Systems, Illustrated For Three Projects ($000 omitted.)

	Project A	Project B	Project C	Indirect	Total
Direct Labor	50	30	10		
Direct Nonlabor	0	20	40		
Overhead on Direct Labor				90	
General & Administrative				30	
Burden					

$$\left.\begin{array}{l}\text{Direct Labor}\\ \text{Direct Nonlabor}\end{array}\right\} = 150 \left\{\begin{array}{l}\text{Labor}\\ \text{Purchases}\\ \text{Travel}\end{array}\right.$$

$$\left.\begin{array}{l} 90 \\ 30 \end{array}\right\} = 120 \left\{\begin{array}{l}\text{Fringe Benefits}\\ \text{Indirect time of direct labor personnel}\\ \text{Indirect labor personnel}\\ \text{Facility costs}\\ \text{General supplies}\\ \text{Publications}\end{array}\right. = 270$$

Method 1

	Project A	Project B	Project C	Indirect	Total
Direct Labor	50	30	10		
Direct Nonlabor	0	20	40		
Direct Total	50	50	50	= 150	
Burdens on Direct Total	40	40	40	= 120	
Total Costs	90	90	90	270	

Method 2

	Project A	Project B	Project C	Indirect	Total
Direct Labor	50	30	10		
Direct Nonlabor	0	20	40		
Direct Total	50	50	50		
Overhead on Direct Labor	50	30	10	= 90	
General & Administrative					
Burden on Direct Total	10	10	10	= 30	
Total Costs	110	90	70	270	

$$\left.\begin{array}{l} = 90 \end{array}\right\} = 150 \qquad \left.\begin{array}{l} = 30 \end{array}\right\} = 120 \qquad = 270$$

TABLE 9-1. *continued*

	Project A	Project B	Project C	Indirect	Total
Method 3					
Direct Labor	50	30	10		
Overhead on Direct Labor	50	30	10	= 90	
Direct Nonlabor	0	20	40		= 120
"Prime" Costs	100	80	60		
General & Administrative					
Burden on "Prime" Costs	12.5	10.	7.5	= 30	
Total Costs	112.5	90	67.5		= 270
Method 4					
Direct Labor	50	30	10		
Overhead on Direct Labor	50	30	10	= 90	
Direct Nonlabor–Purchases	0	20	20		= 120
Material Handling Burden	0	5	5	= 10	
Direct Nonlabor–Other	0	0	20		= 30
"Prime" Costs	100	85	65		
General & Administrative					
Burden on "Prime" Costs	8	6.8	5.2	= 20	
Total Costs	108	91.8	70.8		= 270

cost accounting system. If you are placing a labor intensive contract with a subcontractor, you should not use a subcontractor who practices method 3 as opposed to method 1.

In addition to understanding project budgets, you should understand the relationship between project budgets and administrative budgets, as illustrated in Figure 9-5. The administrative budgets are overhead, general and administrative, and capital. As we saw in Table 9-1, overhead and G and A may be combined into a single burden pool. Capital expenditures enter into the overhead (or G and A) budgets by requiring inclusion of depreciation. Thus, the direct cost of a project, its own direct labor and nonlabor expenses, is not really a measure of its cost to the organization. The project must, in common with all other organizational activities, carry burdens that depend on other organizational activities and budgets.

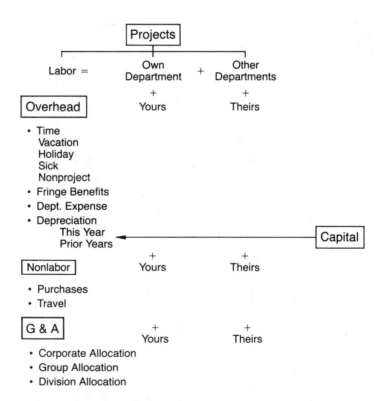

Figure 9-5. Budget relationships.

Source: "Basic Management Skills for Engineers and Scientists." Lecture notes by Milton D. Rosenau, Jr. Copyright © 1982 by the University of Southern California. Reprinted by permission.

TYPICAL PROBLEMS

There are three important problems in planning the cost dimension of the Triple Constraint. First, many project groups or project managers have a deplorable tendency to make cost estimates for support group work. This forecloses the possibility of benefiting from support group expertise and violates the Golden Rule. This is easily solved by requiring every department to approve the estimate for the work it will do.

A second problem is dealing with inflated estimates by support groups. Here the project manager can first try discussion and negotiation. If that does not produce a satisfactory agreement, the project manager could alter the nature of the requested support work. Two other possible solutions are to subcontract the support work to another company or appeal to higher management.

Higher management, if they decide to "buy in," often cause a third problem. If you are convinced that buying in is disaster, you can request that someone else assume project management. Or you can record your objections in a memo and try to accomplish the promised work within the budget. Finally, you can undertake the job and work actively to sell your customer on changes of scope that provide an opportunity for more funding.

MATERIALS STUDY PROJECT

The materials study project consists of ten tasks divided among five departments. Having identified the tasks by the work breakdown structure (Figure 6-4) and the involved departments by reference to the logical work elements and the organizational arrangements, it is then possible to identify which tasks have to be estimated as to cost and schedule by which departments. This is summarized in Figures 9-6 and 9-7. Fig-
continued

Each department should estimate its work on each task.

ure 9-6 shows the twenty-five estimates to be made and how computers can summarize them. The estimates and summaries are illustrated in Figures 9-8 through 9-14 and are identified in Figure 9-7.

The computer can be used to simplify planning by quickly summarizing large amounts of numerical data. For very small projects, this can easily be done by hand and is not sufficient justification for using a computer. Figure 9-10 illustrates the summation for task A, and Figure 9-14, for the chemical group.

Once all the cost estimates are summed by task, it is possible to complete a costed WBS (Figure 9-15) should this be desired.

TASK \ SECTION	PROJECT (MECH.)	CHEM.	ELEC.	RESEARCH	TECH. SUPPORT	ALL
A	E	E				S
B	E	E				S
C	E		E		E	S
D	E		E		E	S
E	E	E	E		E	S
F	E				E	S
G	E			E		S
H	E			E		S
J				E		I
K	E	E	E		E	S
ALL	S	S	S	S	S	S

E = ESTIMATE MADE BY INDICATED DEPARTMENT FOR INDICATED TASK

S = SUM OF COLUMN OR ROW PREPARED BY COMPUTER OR ADMINISTRATIVE-AIDE

I = SUM OF COLUMN OR ROW IS IDENTICAL TO ONLY ENTRY IN COLUMN OR ROW

S = SUM OF ENTIRE PROJECT, PREPARED BY COMPUTER OR ADMINISTRATIVE AIDE

Figure 9-6. Estimation of the materials study project is accomplished by making twenty-five individual estimates.

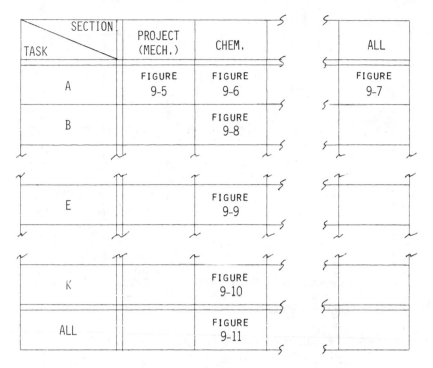

SECTION TASK	PROJECT (MECH.)	CHEM.	ALL
A	FIGURE 9-5	FIGURE 9-6	FIGURE 9-7
B		FIGURE 9-8	
E		FIGURE 9-9	
K		FIGURE 9-10	
ALL		FIGURE 9-11	

Figure 9-7.   Guide to examples of estimates and summaries.

PROJECT _Mat'l Study_   TASK _A- Sel. Mat'ls_   DEPARTMENT _Mech_

COST ELEMENT			HOURS EACH MONTH						TOTALS	
			1	2	3	4	5	6	HOURS	DOLLARS
LABOR	SR. PROF.	$25/hr.	6						6	
	JR. PROF.	$20/hr.								
	SR. TECH.	$15/hr.								
	JR. TECH.	$10/hr.								
			DOLLARS EACH MONTH							
DOLLARS	LABOR COST									
	OVERHEAD	100%								
	DIRECT NONLABOR									
	PRIME COSTS									
	G & A	15%								
	TOTAL COSTS									
	PROFIT	20%								
	TOTAL BILLING									

ASSUMPTIONS _I'll spend 2 hr/wk with Cheryl_

PREPARED BY _Will Chase_   DATE _6 Nov 84_   APPROVED BY _Ed Ross 11/7_

Figure 9-8.   Estimate for task A by the mechanical engineering group.

PROJECT _MATERIALS STUDY_ TASK _A- SELECT MATERIALS_ DEPARTMENT _CHEMICAL_

COST ELEMENT		HOURS EACH MONTH						TOTALS	
		1	2	3	4	5	6	HOURS	DOLLARS
LABOR	SR. PROF. $25/hr.	60						60	
	JR. PROF. $20/hr.								
	SR. TECH. $15/hr.								
	JR. TECH. $10/hr.								
		DOLLARS EACH MONTH							
DOLLARS	LABOR COST								
	OVERHEAD 100%								
	DIRECT NONLABOR								
	PRIME COSTS								
	G & A 15%								
	TOTAL COSTS								
	PROFIT 20%								
	TOTAL BILLING								

ASSUMPTIONS _3 weeks, half-time of CW_

PREPARED BY _CHERYL WILLIAMS_   DATE _Nov 5, 1984_   APPROVED BY _Ed Ross 11/6_

Figure 9-9.   Estimate for task A by the chemical group.

PROJECT___MATERIALS STUDY___ TASK _A-SELECT MATERIALS_ DEPARTMENT ___ENGINEERING___

COST ELEMENT		HOURS EACH MONTH						TOTALS	
		1	2	3	4	5	6	HOURS	DOLLARS
LABOR	SR. PROF. $25/hr.	66						66	1,650
	JR. PROF. $20/hr.								
	SR. TECH. $15/hr.								
	JR. TECH. $10/hr.								
		DOLLARS EACH MONTH							
DOLLARS	LABOR COST	1,650							1,650
	OVERHEAD 100%	1,650							1,650
	DIRECT NONLABOR	0							0
	PRIME COSTS	3,300							3,300
	G & A 15%	495							495
	TOTAL COSTS	3,795							3,795
	PROFIT 20%	759							759
	TOTAL BILLING	4,554							4,554

ASSUMPTIONS _____

PREPARED BY _____   DATE ___19 NOV 84___   APPROVED BY_____

Figure 9-10.   Summary of task A work in the engineering section (sum of mechanical engineering and chemical group efforts), prepared by computer or an administrator.

PROJECT _MATERIALS STUDY_   TASK _B-OBTAIN MATERIALS_ DEPARTMENT _CHEMICAL_

COST ELEMENT			HOURS EACH MONTH						TOTALS	
			1	2	3	4	5	6	HOURS	DOLLARS
LABOR	SR. PROF.	$25/hr.	40	4	2				46	
	JR. PROF.	$20/hr.								
	SR. TECH.	$15/hr.								
	JR. TECH.	$10/hr.								
			DOLLARS EACH MONTH							
DOLLARS	LABOR COST									
	OVERHEAD	100%								
	DIRECT NONLABOR		4,400							4,400
	PRIME COSTS									
	G & A	15%								
	TOTAL COSTS									
	PROFIT	20%								
	TOTAL BILLING									

ASSUMPTIONS _CW full-time at supplier during first week + $400 travel + $4,000 subcontract; 1 hr/wk during next 4 wks to answer questions; 2 hr last week._

PREPARED BY _CHERYL WILLIAMS_   DATE _Nov 5, 1984_   APPROVED BY _Ed Ross 11/6_

Figure 9-11. Estimate for task B by the chemical group.

PROJECT _MATERIALS STUDY_   TASK _E- EXPERIMENT_ DEPARTMENT _CHEMICAL_

COST ELEMENT			HOURS EACH MONTH						TOTALS	
			1	2	3	4	5	6	HOURS	DOLLARS
LABOR	SR. PROF.	$25/hr.				60	80	80	220	
	JR. PROF.	$20/hr.								
	SR. TECH.	$15/hr.								
	JR. TECH.	$10/hr.								
			DOLLARS EACH MONTH							
DOLLARS	LABOR COST									
	OVERHEAD	100%								
	DIRECT NONLABOR									
	PRIME COSTS									
	G & A	15%								
	TOTAL COSTS									
	PROFIT	20%								
	TOTAL BILLING									

ASSUMPTIONS _11 weeks, half-time of CW_

PREPARED BY _CHERYL WILLIAMS_ DATE _Nov 5, 1984_ APPROVED BY _Ed Ross 11/6_

Figure 9-12. Estimate for task E by the chemical group.

PROJECT **MATERIALS STUDY**   TASK **K- FINAL REPORT**   DEPARTMENT **CHEMICAL**

COST ELEMENT		HOURS EACH MONTH						TOTALS	
		1	2	3	4	5	6	HOURS	DOLLARS
LABOR	SR. PROF. $25/hr.						20	20	
	JR. PROF. $20/hr.								
	SR. TECH. $15/hr.								
	JR. TECH. $10/hr.								
		DOLLARS EACH MONTH							
DOLLARS	LABOR COST								
	OVERHEAD 100%								
	DIRECT NONLABOR								
	PRIME COSTS								
	G & A 15%								
	TOTAL COSTS								
	PROFIT 20%								
	TOTAL BILLING								

ASSUMPTIONS _1 week, half-time of CW_

PREPARED BY _CHERYL WILLIAMS_ DATE _Nov 5, 1984_   APPROVED BY _Ed Ross 11/6_

Figure 9-13.   Estimate for task K by the chemical group.

PROJECT _____ MATERIALS STUDY _____   TASK _____ ALL _____ DEPARTMENT _____ CHEMICAL

COST ELEMENT		HOURS EACH MONTH						TOTALS	
		1	2	3	4	5	6	HOURS	DOLLARS
LABOR	SR. PROF. $25/hr.	100	4	2	60	80	100	346	8,650
	JR. PROF. $20/hr.								
	SR. TECH. $15/hr.								
	JR. TECH. $10/hr.								
		DOLLARS EACH MONTH							
DOLLARS	LABOR COST	2,500	100	50	1,500	2,000	2,500		8,650
	OVERHEAD 100%	2,500	100	50	1,500	2,000	2,500		8,650
	DIRECT NONLABOR	4,400							4,400
	PRIME COSTS	9,400	200	100	3,000	4,000	5,000		21,700
	G & A 15%	1,410	30	15	450	600	750		3,255
	TOTAL COSTS	10,810	230	115	3,450	4,600	5,750		24,955
	PROFIT 20%	2,162	46	23	690	920	1,150		4,991
	TOTAL BILLING	12,972	276	138	4,140	5,520	6,900		29,946

ASSUMPTIONS _____

PREPARED BY _____   DATE _19 NOV 79_ _____ APPROVED BY _____

Figure 9-14.   Summary for entire materials study of work to be done by the chemical group, prepared by computer or administrator.

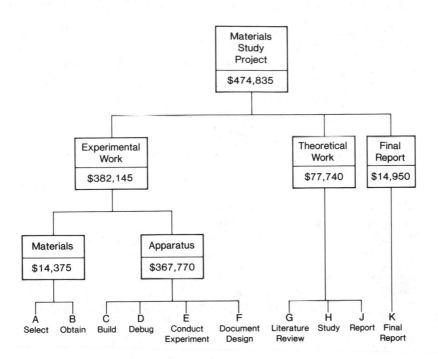

Figure 9-15.   Costed WBS for materials study project.

Cost estimates are usually made in dollars.

Cost estimates can be made top down or bottom up, but a better method is to do both, in that order.

A parametric cost estimate may be useful and simple, especially if your project is reasonably similar to others for which historical data are available.

The elements of a project cost accounting system, a means to tally costs by project and project task, are labor, overhead burden, nonlabor, and general and administrative burden.

FURTHER READING

R. H. Clough and G. A. Sears. *Construction Project Management,* 2nd ed. New York: Wiley-Interscience, 1979.
   Chapter 3, although specific to the construction industry, is a thorough treatment of estimating.

V. G. Hajek. *Management of Engineering Projects.* New York: McGraw-Hill, 1977.

Chapter 6 has a brief discussion of cost estimating pitfalls.

J. A. Maciariello. *Program-Management Control Systems.* New York: Wiley-Interscience, 1978.

Chapter 7 is a brief treatment of project cost accounting systems.

"PRICE Parametric Price Models." Cherry Hill, NJ: RCA PRICE Systems.

These are descriptive brochures on all the RCA PRICE models.

M. D. Rosenau. *Successful Project Management.* Belmont, CA: Lifetime Learning Publications, 1981.

Chapter 9, pages 99–114, contains many examples of reports from a computer-based project planning system.

M. Spinner. *Elements of Project Management: Plan, Schedule, and Control.* Englewood Cliffs, NJ: Prentice-Hall, 1981.

Chapter 9, pages 161–184, provides a thorough description of the PERT cost system.

The Impact of Limited Resources

This chapter deals with the impact of finite resources on project plans. This topic typically involves two or three dimensions of the Triple Constraint. First I discuss resource allocation and how to resolve resource constraints. Then I present techniques that allow analysis of schedule and budget trade-offs.

RESOURCES

Resources are either people or things. Human resources may include everyone in a particular organizational unit or those with a specific skill (optical design, analytic chemistry, mechanical design engineering, or electronic circuit design, for example). Things include any kind of equipment, for instance, lathe availability or pilot plant time as well as floor space to house the equipment and the people. Money may also be considered a nonhuman resource.

Allocation

There are three reasons to consider resource allocation in an engineering project management environment. First, forecasted use of some key resource (for instance, circuit designers) may indicate there will be surplus personnel at some

*Surplus resources
waste money and
talent.*
future period. This information should warn the appropriate
managers either to obtain new business to utilize the surplus
talent or to plan to reassign the involved personnel.

Another reason for resource allocation is to avoid inherent
inconsistencies, for instance, using a particular resource
(Jane Draftsperson, for example) on two tasks at the same
time. Preparing a network diagram to a time base emphasizes
resource allocation and reveals latent conflicts.

A network diagram can show what resources are required
and when, which may reveal that more of some resources will
be needed than will be available at some time. When you
discover this, you must adjust the network diagram to shift
the overloaded resource requirement to some other time. If
you fail to do this, slippage will occur. Figure 10-1 illustrates
resource allocation. In this case, the resource is the personnel
headcount. Tasks A and B, each of eight weeks duration,
require three and five personnel, respectively. Tasks C, D, and
E are not on the critical path, and examination of the earliest
and latest times for them shows they can be commenced
immediately or as late as the eleventh week. If the company
performing this project employs only six people, task D

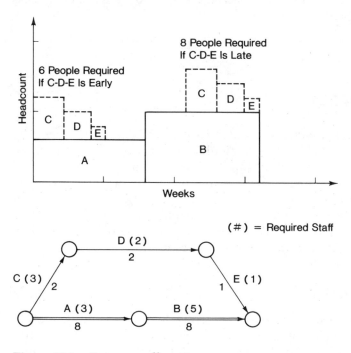

Figure 10-1.   Resource allocation.

would have to start early enough to be completed before the end of the eighth week, when task B is scheduled to start. If task D starts later than the start of the sixth week, some (or all) of task D will be scheduled in parallel with task B, requiring seven personnel.

A third use of this kind of analysis occurs in a large company. Imagine that tasks C, D, and E are performed by a particular support department, for instance, the design and drafting section. If the design and drafting section was provided with resource allocation information for all projects, as shown in Figure 10-1, they could identify the earliest and latest dates at which the support, in this case, C, D, and E, would have to be applied. Doing the same for all projects would allow the support group to even out its work load and to identify in each case the impact of any slippage.

Removing Resource Constraints

Consider the CPM network diagram in Figure 10-2. After planning the work, you arrive at a summary of labor skills required for this project (Table 10-1). Suppose that you have only nine junior engineers. What are your options? What might you do? What are the risks?

Figure 10-3 graphically summarizes the junior engineer staffing each week. There are two weeks when ten junior engineers are required, but the constraint is that only nine are available. Some of your options are to use overtime, use senior engineers to do the work, or delay task F (which has slack) by two weeks. The latter, which seems most appropriate, runs the risk of reducing slack that may be needed later. In general, you can do the following to remove resource constraints when only one project is involved:

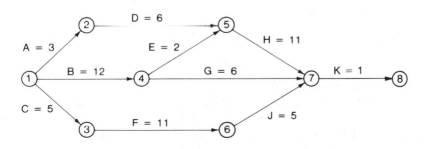

Figure 10-2. CPM diagram.

TABLE 10-1. Resource Allocation, Using Network of Figure 10-2.

| Task | Planned Duration | Required Resources | | | | | |
		Senior Eng'r	Junior Eng'r	Design	Mech	Elec	Inspect
A	3	2	4	4			
B	6		2			3	
C	12*	3	4	4			
D	2*		2		5		
E	11*	2	5		2	2	
F	6		2	7			
G	6	4	1				
H	10	1	3				
J	5		2				
K	1*	2	2				3

* = Task is on critical path.

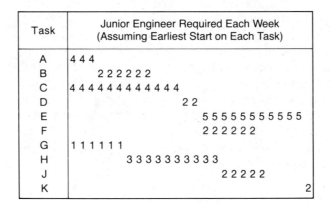

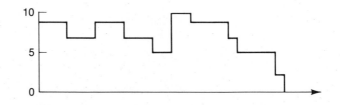

Figure 10-3. Junior engineer resource requirement on first project network diagram of Figure 10-2.

1. Identify resource requirements for tasks on the critical path.

2. Add resource requirements for other tasks, using desired start dates.

3. Compare resource requirements with resource availability.

4. Identify options to remove resource conflicts that are found. You may verify that the conflict is real, adjust start dates for tasks with slack, give the schedule more attention to reduce downtime inefficiency, improve productivity (with new tools, improved match of people with tasks, or incentives), adjust resource availability temporarily (by planning overtime, rescheduling vacations, obtaining temporary employees, or subcontracting tasks), change resource availability (by training or promoting surplus skill to fill needs or hiring new people), change the time schedule (accelerate the critical path, which will put you ahead of schedule, delay the critical path, which will make you late, or change the start date of new projects), or change the plan (specifications, task sequence, or standards).

A project schedule that requires use of already obligated resources is unrealistic.

Suppose that you have a simultaneous second project (Figure 10-4). Suppose that task W requires four inspectors and you have only six. That's no problem on the second project taken alone, but it is when you try to accommodate its need for inspectors with the first project's need, as shown in

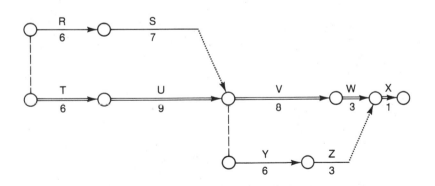

Figure 10-4. Resource allocation with second project. (This project starts at the same time as the first project. The network diagram, including task durations, is shown.)

Table 10-1. Both task W in the second project and Task K in the first project are on the critical paths, and they need a total of seven inspectors at the same time. Something has to be done.

The procedure in this case is the same as removing conflicts on a single project, except that:

1. Projects must be ranked by priority.

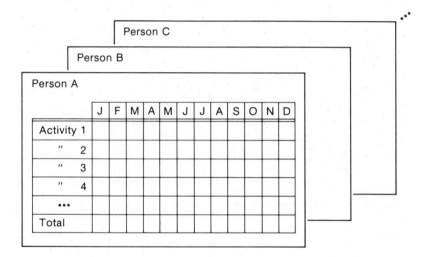

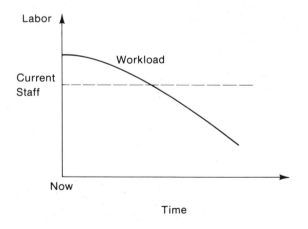

Figure 10-5. Work load forecasting and planning.

Source: "Basic Management Skills for Engineers and Scientists." Lecture notes by Milton D. Rosenau, Jr. Copyright © 1982 by the University of Southern California. Reprinted by permission.

2. The highest priority project gets the first claim on available resources.

3. The second priority project gets second claim, and so on.

The result is that lower priority projects usually get delayed, performance is compromised, or there is a lot of (unplanned) subcontracting. Sometimes it is cost-effective to accelerate a small or low priority project to get it out of the way and thus avoid a major conflict that would otherwise arise.

Lower priority projects usually lose the competition for limited resources.

Project schedules are usually prepared initially without regard to whether the required resources will actually be available when desired or required. Thus, there can be a serious problem if the impact of resource constraints is overlooked. The first step to avoid this problem is to refine the schedule for your project so that all tasks are consistent with available resources. Then the resource requirements of other projects must be checked and conflicts resolved. These other projects include both those that might start and be in process during your project and those existing (or planned) projects that are supposed to be completed before your project starts but that are delayed to impact your project. If this is not done, the lack of resources will not magically cure itself; it will become an obstacle when there is less (or no) time to devise an alternative schedule.

Resource allocation can be done by inspection of predicted resource use. The top of Figure 10-5 illustrates forms that could be used to summarize individual work loads, the total summation of which typically appears as in the bottom of the figure. Computers can also do resource leveling and work load prediction and are not prone to making arithmetical mistakes (assuming, of course, they have been programmed correctly and the data have been loaded accurately). Microcomputers can also be used. Figures 10-6 and 10-7, prepared easily using a spread sheet program, show how microcomputers can quickly assist with work load forecasting.

A network diagram is not merely a schedule dimension plan; it also clarifies resource allocation.

TIME VERSUS COST TRADE-OFF

CPM has historically been associated with network diagrams in which there is considered to be a controllable time for each activity. This implies that activities can be accelerated by devoting more resources to them. Thus, there is a time versus

	Jan	Feb	Mar	Apr
Project A				
Senior	1	2	3	4
Intermediate	5	6	7	8
Junior	1	2	4	8
Project B				
Senior	4	3	2	1
Intermediate	2	4	6	8
Junior	3	6	9	9
Project C				
Senior	3	4	5	6
Intermediate	3	5	7	9
Junior	1	3	5	7
Total				
Senior	8	9	10	11
Intermediate	10	15	20	25
Junior	5	11	18	24
Available Resources				
Senior	8	8	8	9
Intermediate	11	13	15	15
Junior	6	8	10	12
Resource Surplus (+) or Deficit (−)				
Senior	0	−1	−2	−2
Intermediate	1	−2	−5	−10
Junior	1	−3	−8	−12

Figure 10-6. Original work load for three projects.

cost trade-off for each activity and consequently for a path or the entire project.

Figure 10-8 shows this kind of situation. If you are trying to accelerate a project, you should accelerate the critical path. Of all the activities on the critical path, the most economical to accelerate are those with the lowest cost per amount of time gained.

Consider the following situation that you might face as the project manager for the engineering project illustrated in Figure 8-13:

	Jan	Feb	Mar	Apr
Project A				
Senior	1	2	3	4
Intermediate	5	6	7	8
Junior	1	2	4	8
Project B				
Senior			4	3
Intermediate			2	4
Junior			3	6
Project C				
Senior	3	4	5	6
Intermediate	3	5	7	9
Junior	1	3	5	7
Total				
Senior	4	6	12	13
Intermediate	8	11	16	21
Junior	2	5	12	21
Available Resources				
Senior	8	8	8	9
Intermediate	11	13	15	15
Junior	6	8	10	12
Resource Surplus (+) or Deficit (−)				
Senior	4	2	−4	−4
Intermediate	3	2	−1	−6
Junior	4	3	−2	−9

Figure 10-7.   Revised work load for three projects when project B is delayed two months.

   To carry out this project (assuming all times are in weeks), you must rent a standby electrical generator for the entire duration of the project (however long it is) at a cost of $1,000 per week. Your purchasing department has told you that the subcontractor performing task B has offered to shorten its performance time by as much as five weeks (that is, to seven weeks) but will charge $800 per week for every week less than the original twelve weeks (that is, a premium charge of $800 for eleven weeks delivery, $1,600 for ten weeks delivery, or $4,000 for seven weeks delivery). You can save $200 per week by accepting the subcontractor's offer.

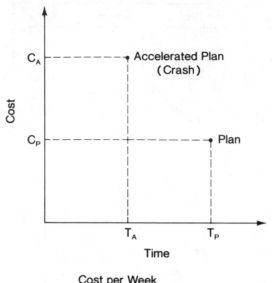

$$\text{Cost per Week}\atop\text{(Day, Month)}\atop\text{to}\atop\text{Accelerate} = \frac{C_A - C_P}{T_P - T_A}$$

Figure 10-8.   CPM time versus cost trade-off.

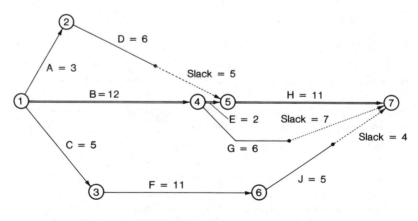

The saving of $200/week can be obtained
for only four weeks, after which path
C-F-J also becomes a critical path.

Figure 10-9.   Time-oriented CPM diagram showing that the saving
of $200/week can be obtained for only four weeks, after which path
C-F-J also becomes a critical path. (Depending on the certainty you
feel for C-F-J and B-E-H, it might be better to shorten only activity
B by three weeks, thus maintaining only one critical path.)

However, you have to look at your network diagram before rushing to accept the offer (Figure 10-9).

As you can see, once task B is shortened from twelve to eight weeks (at a cost premium of $3,200, producing a $4,000 saving on the standby generator), there is a second critical path. Thus, you cannot advantageously shorten task B by five weeks, only by four. In fact, you might prefer to shorten task B by only three weeks to avoid having two critical paths.

Figure 10-10 shows another aspect of this. The direct cost curve depicts those costs associated with carrying out the project that are time dependent and for which there is a cost premium associated with shortening the program. In addition, there might very well be continuing costs associated with the program, for instance, the rental of standby power generators or other such facilities. In this kind of situation, there will be a time that leads to the lowest cost for the project.

TYPICAL PROBLEMS

A major resource problem occurs when project schedules change. This may be caused by a change in your project or another project or by a shift in the start date
continued

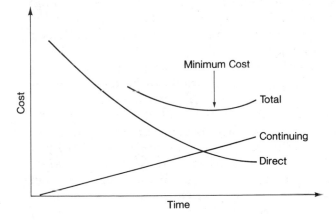

Figure 10-10. Finding the lowest cost.

of a project. As I discussed in Chapter 3, you can never be certain which proposals will be approved or when they will start. Therefore, you must be constantly alert for these changes and their impact on the resources you plan to use.

Computers are very helpful for manipulating and summing personnel resource requirements on many simultaneous projects and thus can help alert you to prospective resource constraint problems. When computers are used to assist with planning (or anything, for that matter), there is always the danger of entering incorrect data or making programming errors. With standard project control software, the software is proven, but there can still be incorrect data entry. To avoid this kind of problem, verify data entry or run manual spot checks of output. Never blindly accept any computer output as gospel.

MATERIALS STUDY PROJECT

The materials study project can be used to illustrate a resource allocation problem. Cheryl Williams is slated to work on three projects and also plans to attend a society meeting. A work load forecast for her (Figure 10-11) would show the conflict. Once it is identified, corrective action could be initiated (such as planning overtime work, canceling her society participation, or finding some other chemist to work on the quill development project).

In more complex situations, where there are many projects, holidays, and vacations to be considered, a computer is usually better able to handle the data. Project control software often includes priority scheduling in which low priority projects are rescheduled so as not to conflict with high priority projects (which usually results in low priority projects being late). "What if" capability is also often included, allowing managers to explore all the consequences of various scheduling options.

Month & Hours Allowed / Task or Project	1	2	3	4	5	6
	160	160	200	160	160	200
Materials Study	100	4	2	60	80	100
Ink Development	20	60	80	60	40	
Quill Development			40	40	40	
Chemical Society				40		
Total Hours	120	64	122	200	160	100
Available or ⟨Conflict⟩	40	96	78	⟨40⟩	—	100

Figure 10-11. Projected six-month work load for individual, showing conflict in month 4.

HIGHLIGHTS

Resources, whether people or things, should be carefully allocated in a project.

A network diagram can clarify resource allocation.

Each activity, critical path, and project has a time versus cost trade-off.

Computers can aid project planning in several ways, although care must be taken to avoid entering incorrect data and making programming errors.

FURTHER READING

R. H. Clough and G. A. Sears. *Construction Project Management*, 2nd ed. New York: Wiley-Interscience, 1979.
  Chapter 7 is a thorough treatment of the time reduction trade-off, using construction industry examples.

C. F. Gray. *Essentials of Project Management*. Princeton, NJ: Petrocelli, 1981.
  Chapter 10, pages 97–140, provides a thorough discussion of resource scheduling.

J. A. Maciariello. *Program-Management Control Systems*. New York: Wiley-Interscience, 1978.
  Chapter 8 is a good but brief treatment of resource allocation.

J. Mulvaney. *Analysis Bar Charting—A Simplified Critical Path Analysis Technique,* U.S. ed. Washington, D.C.: Management Planning Systems, 1977.

Chapter 6 has some discussion about the impact of resources on schedule.

Project Management Institute. *Survey of Project Management Software Packages.* Drexel Hill, PA: Project Management Institute, October 1982.

This publication contains the most complete listing of currently available scheduling and control software. Note, however, that it omits some important existing products and that new project management software for microcomputers appears frequently.

11

Contingency and Risk

In this chapter, I review methods to provide vitally important schedule and cost contingency and then discuss the trade-off with risk. Finally, I discuss a hybrid system, C/SCSC, used in large Department of Defense contracts.

CONTINGENCY

Plans represent the future. Because nobody has a crystal ball, plans must include contingency. In fact, this contingency should be placed on each of the three dimensions of the Triple Constraint. Smart customers know that contingency is required and often want to know how much and what kind is included—not to try to eliminate it, but rather to be assured that their own plans (which include your project) can be achieved. (Unfortunately, some government procurement regulations prohibit or inhibit this realistic approach, which may contribute to the costly overruns frequently reported.)

On the performance dimension, it is important that the contingency not take the form of gold plating. Where appropriate, include a small design margin. For instance, if the goal is to have a new car's weight be only twenty-five hundred pounds, it might be appropriate to try to design the car to weigh twenty-three hundred pounds. However, never carry this to an extreme (targeting the car's weight at fifteen hundred pounds, for example).

Build contingency into all project plans.

Contingency is most often associated with the schedule and cost dimensions because projects will inevitably encounter difficulties there. Many things that occur simply require more time and money than planners think. For instance, whenever you must interact with other people, obtaining your boss' approval, perhaps, their schedules constrain you. You will not have instant access, and a delay will occur. Customer furnished items, such as interface hardware, frequently do not arrive when expected or in the condition promised. Work done at remote locations often takes longer than work done in your own organization's facilities because things like copying machines or other support resources are simply not as conveniently available. If your project involves hiring people, it takes time to train them and make them effective workers on the job. Even if you don't have to hire people, you must still indoctrinate the people who are assigned to work on your project. Similarly, there will be illnesses and vacations pulling people away from the job. These and other tasks, some of which are included in the following list, make it important to build in schedule and cost contingency.

Resource conflicts

Interface with others

Get approvals

Get support from other groups

Place major subcontracts and purchase orders

Make mistakes

Train people

Replace sick and vacationing personnel

Obtain security clearances

Obtain customer furnished equipment

Work at remote locations

Cope with travel delays

Handle customs duty clearance

Adjust for labor strikes

Comply with customer procedures

Advance the state of the art

Accommodate computer downtime

There are several ways to insert time and cost contingency. The first is to have everybody who provides an estimate make his or her own time and cost contingency estimates.

The problem with this approach is that contingency gets applied on top of contingency, which is then applied on top of other contingencies, and so on. It does not take many multiplications of 110 or 120 percent before the price of the entire project exceeds the customer's reach, and it does not take many extra hours, days, weeks, or months before the schedule becomes unreasonable.

The second method to insert schedule and cost contingency is to put a small amount of contingency, 5 or 10 percent, on each activity in the network. This method is fine, but it misses the point that some activities can be accurately estimated and some others cannot.

A third method to add contingency is to add an unplanned (and, it is hoped, unrequired) activity. For instance, adding a system management task for the duration of the project is really a variation of the second method but provides only cost contingency. A given level of cost contingency, the amount of the system management task, is attached to the entire project. A weakness of this method is that it does not inherently provide a time cushion as well.

A fourth method is a variation on the third method. This fourth method, which I prefer, is to add one or more additional plausible (but, it is again hoped, unrequired) tasks near the end of the project, as shown in the middle of Figure 11-1. This has the effect of pushing activities forward from the project's scheduled completion, in this case, shipment, to the earliest possible point. These extra tasks thus have the effect of providing schedule contingency. Because money is devoted to these extra tasks (not necessarily in direct proportion to their increased time), they also supply a cost contingency. However, both the time and money contingencies are purely arbitrary.

A fifth method is a variation of the second method (that is, adding 5 or 10 percent). The variation, which is far better than the second one, is to explain to everyone providing estimates that they should be as accurate (or optimistic or pessimistic) as possible. Then the entire group can discuss how much schedule and cost contingency should be put on which activities. This can be done by considering the likelihood of things going wrong, the importance of such an outcome, and the maximum and minimum impacts produced by this undesired outcome. There might be some highly uncertain activities that receive a contingency of 50 percent or more. Conversely, a final report might be assigned no or only a small percentage contingency. This fifth method is illustrated in the bottom of Figure 11-1.

Beware of applying contingency upon contingency.

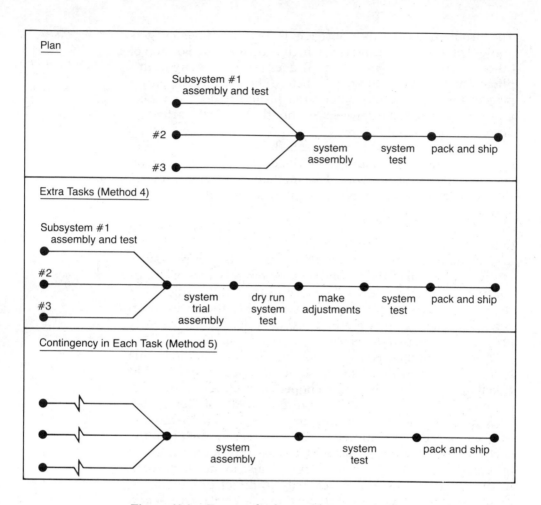

Figure 11-1.   Two methods to add time and cost contingency.

Source: Based on illustration from "Project Management." Lecture notes by Milton D. Rosenau, Jr. Copyright © 1981 by the Association for Media-based Continuing Education for Engineers, Inc. (AMCEE). Reprinted by permission.

RISK

In fact, multiple time (and cost) estimates can be made for each task. Then the project manager and task managers can decide which estimate to present in selling the project. This is always a trade-off with risk (see Figure 11-2). The shorter the proposed time schedule (or the lower the bid price), the greater the risk of time (or cost) overrun.

Risk actually has two components: tangible (for instance, financial) and psychological. The latter is often more impor-

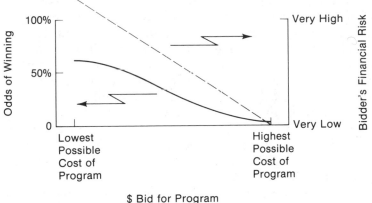

Figure 11-2. Contingency and risk trade-off.

tant because some people have a very low tolerance for tangible risk and thus insist on padding their schedule and budget estimates. Many engineers, educated in the precision of numbers, hate to make a numerical estimate (time or money) that may later be proven wrong; therefore, they have a low psychological tolerance for making optimistic time and cost estimates. Such a psychological bias induces engineers to make pessimistic time and cost estimates so that they can then do better. (However, this bias does not prevent naivete from producing optimistic estimates.)

Risk has both tangible and psychological components.

A more senior manager has a different perspective. He or she is likely to have several project or task manager subordinates. Thus, the senior manager can average his or her risk over several projects or tasks. That is, one project or task may overrun, but there is also likely to be one with a compensating underrun.

In general, when two people are in a dispute or argument about the schedule or budget for a task or project, the point of contention is not the estimate per se; it is about the risk. Unfortunately, the energy of the argument is usually concentrated on whether the work should take, for example, four or six weeks (or cost $4,000 or $6,000), which cannot be known with certainty in advance. It would be far more productive to discuss what the risks are if the proposed schedule is planned for four weeks rather than for six weeks.

Some computer programs can control graphic output terminals to produce network diagrams or bar chart displays. Computers can also furnish summary project planning and

Computers are often useful for contingency planning.

control information for several projects and can easily provide a wide variety of other planning detail to the project manager, support group managers, and task leaders. Thus, it is easily possible to make several different time and cost estimates for each project task, where different amounts of contingency and risk are involved. This can provide a summary overview of the project under these different assumptions. The resulting computer output then allows a really useful discussion. Rather than have a senior manager (or customer) say the project proposal is too long or too expensive, they can examine the greater risks attendant to settling on a shorter schedule or lower budget.

Project planning and control software programs for microcomputers are also becoming widely available. Compared to the software that runs on minicomputers and mainframe computers, these programs are not as flexible or fast. Nevertheless, they are adequate in many cases.

COST/SCHEDULE CONTROL SYSTEM CRITERIA

The Cost/Schedule Control System Criteria (C/SCSC) was devised by the U.S. Department of Defense to predict cost overruns early in major military procurement projects and thus reduce risk. On many projects, the actual cost being reported conformed closely to the initial planned cost, at least during the initial months, but this concealed the fact that actual work accomplishments were lagging behind plan. Consequently, the total project cost for the work performed exceeded planned cost for the work accomplished. Thus, when projects were far advanced and major commitments had been made to them, it became apparent that there were going to be major cost overrun problems at the end for which no contingency reserve existed.

As with many things done by the military, this is a very formal, rigid system. It provides a standard terminology and approach, which clearly are helpful, but it saddles both the military and its contractors with a highly detailed system.

C/SCSC is used on major military projects.

C/SCSC must be initiated in the planning phase, but it is really more useful as a control tool than as a planning device.

To work with this system, you must deal with three quantities, shown in Figure 11-3. The budgeted cost of work scheduled (BCWS) is compared with the budgeted cost of work performed (BCWP) in a given period. The difference between

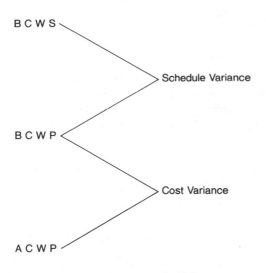

Figure 11-3. C/SCSC methodology.

these is a measure of schedule variance. In addition, the actual cost of work performed (ACWP) is compared to the budgeted cost of the work performed, and the difference, if any, is a cost variance. The details of the C/SCSC system are far too intricate to explain here and are of only limited interest. If you must use it, seek specialized training.

TYPICAL PROBLEMS

The worst problem concerning contingency is that some managers and customers believe they can save time and money by eliminating it. This belief, or pretense, that contingency is not required restricts the project manager's ability to cope quickly, efficiently, and effectively with the problems that inevitably arise on a project. The lack of contingency reserve (especially time and money) means that the project manager must have a customer or management negotiation every time a problem arises.

MATERIALS STUDY PROJECT

Table 11-1 illustrates the different time estimates that might have been made under different risk assumptions. Note that the critical path is changed when the pessimistic estimates are used because tasks A and B are much longer.

TABLE 11-1.   Multiple Time (and Cost) Estimates, Using Network Diagram of Figure 10-2.

Task	80% Optimistic	50-50 Plan	20% Pessimistic
A	2	3	6*
B	5	6	14*
C	11*	12*	14
D	2*	2*	4
E	10*	11*	13*
F	5	6	8
G	6	6	8
H	10	10	12
J	5	5	6
K	1*	1*	2*
TOTAL	24	26	35

* = Task is on critical path.

HIGHLIGHTS   *All projects should contain contingency, which may be best inserted by adding tasks near the end or by distributing it in each task.*

The amount of contingency is a trade-off with risk.

FURTHER READING

R. D. Archibald. *Managing High-Technology Programs and Projects.* New York: Wiley-Interscience, 1976.
    Chapter 7 has a very brief discussion of contingency, which Archibald calls management reserves.

Cost/Schedule Control Systems Criteria-Joint Implementation Guide. Published by Departments of the Air Force (AFSC/AFLC Pamphlet 173-5), Army (DARCOM-P 715-5), Navy (NAV-MAT P5240), and Defense Supply Agency (DSAH 8315.2), October 1, 1976.

Cost/Schedule Management of Non-Major Contracts (C/SSR Joint Guide). Published by Departments of the Army (Pamphlet DARCOM-P 715-13), Navy (NAVMAT P5244), Air Force (AFLCP 173-2 and AFSCP 173-3), and Defense Logistics Agency (DLAH 8315.3), November 1, 1978.
    These two books provide an overview of C/SCSC.

A. M. Ruskin and W. E. Estes. *What Every Engineer Should Know About Project Management.* New York: Marcel Dekker, 1982.
    This book has much material about contingency and reserves, but it appears in bits and pieces, rather than in one unified discussion.

Part 3

LEADING THE PEOPLE WHO WORK ON AN ENGINEERING PROJECT

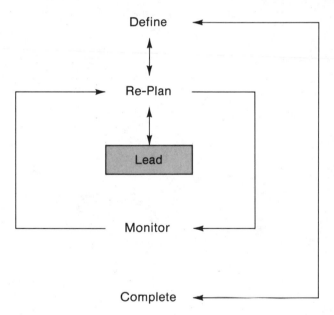

How to Organize an Engineering Project

In this chapter, I describe the three main forms, namely, functional, project, and matrix, by which organizations arrange their internal reporting relationships and chains of command. Any of these three forms may also be used for a project within an organization. Then I describe three other forms by which projects may be managed. Finally, I discuss the informal organization.

THREE PRINCIPAL ORGANIZATIONAL FORMS

Projects have a finite life, from initiation to completion. Conversely, a company, government department, or other organization expects to exist indefinitely. This temporal difference makes it difficult to organize and manage an engineering project within a larger organizational entity.

In addition, engineering projects frequently require the part-time use of resources, but permanent organizations try to use resources full-time. Typical project requirements include the following: one hour of an experimental machinist each day for a week; use of a backhoe next Tuesday for the afternoon; one-quarter of Jane Draftperson's time this month and three-quarters of her time next month; use of Joe Technician full-time as soon as the project's circuit designer completes the design. No economically viable organization can

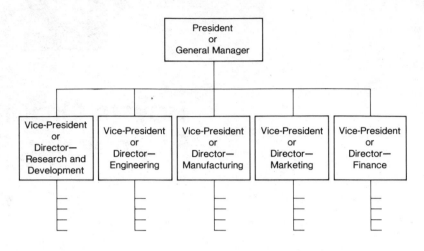

Figure 12-1. Typical organization chart of a functional organization.

afford to stockpile these resources to serve the engineering project's needs instantly. Thus, it is important to organize for project work in adequately responsive ways, and it is important for project managers to recognize that this is a compromise that is not fully responsive to project needs.

Although no organizational form is perfect, it is important to recognize the existence of engineering projects when they are present. This means the organization must plan to accommodate this temporary disturbance and accept some disharmony.

Organizational forms differ in response to projects.

There is a variety of ways that companies or their divisions or governmental organizations can be organized and effectively manage projects. The three most common of these organizational forms are functional, project, and matrix.

Functional

Functional organizations (Figure 12-1) are common in companies dominated by marketing or manufacturing departments (whenever there is a large amount of repetitive work) and exist in other kinds of companies as well. The person asked to manage an engineering project in a company with a functional organization has generally been oriented and loyal to the functional group to which he or she belongs. Specialists are grouped by function, encouraging the sharing of experience and knowledge within the discipline. This favors a continuity and professional expertise in each functional area.

Because such an organization is dedicated to perpetuating the existing functional groups, however, it can be difficult for a project to cross functional lines and obtain required resources. It is not uncommon for hostility to exist between different functions; that is, there are barriers to horizontal information flow, and open channels tend to be vertical, within each function. Absence of a project focal point may trouble a customer interested in understanding the project's status, and functional emphasis and loyalties may impede completion.

From a project management point of view, the functional organization is least desirable.

However, if all the project's required personnel resources (including the project manager) are located within a single functional group, many of the problems can be avoided. Thus, when all project resources are located in one group, the functional organization may be a good choice.

Project

A project organization (Figure 12-2) emerges from a functional organization when the latter impedes project needs. The solution is to move many of the people working on the engineering project from their functional group to the engineering project manager. Line authority for the project is

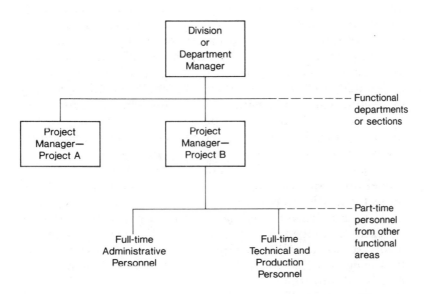

Figure 12-2. Typical organization chart of a project organization.

clearly designated, providing a single focal point for project management. All full-time personnel are formally assigned to the project, thus assuring continuity and expertise.

A major difficulty with this kind of organization is the uncertainty these people feel about where they will go when the project is completed. There is also a tendency to retain assigned personnel too long. In addition, it is a rare project that actually has all the required resources assigned to it. Thus, such an organization still requires the engineering project manager to negotiate with the remaining functional organization for much of the required support.

The project organization form is most useful on large projects of long duration.

If the organization develops additional engineering projects, managing them in this way leads to a splintering, with many separate project centers existing apart from the functional organization. Duplication of facilities and personnel can result. Managers within the functional organization may feel threatened as people are removed from their functional group. This produces another series of stressors. Project organization often inhibits the development of professional expertise in functional specialties and may not effectively utilize part-time assistance from them.

Matrix

The matrix organization (Figure 12-3) is a hybrid that may emerge in response to the pressures resulting from inadequacies with a functional or a project organization. It attempts to achieve the best of both worlds, recognizing the virtues of having functional groups but also recognizing the need to have a specific focal point and management function for each project. Line authority for the project is clearly designated, providing a single focal point. Specialists, including engineering project managers, are grouped by function, encouraging the sharing of experience and knowledge within the discipline. This favors a continuity and professional expertise in each functional area. The matrix organization recognizes that both full-time and part-time assignment of personnel are required and simplify allocation and shifting of project priorities in response to management needs.

The matrix is probably the best organizational option if you have many projects.

The main drawback is that a matrix organization requires an extra management function (namely, project management); so it is usually too expensive for a small organization. It is even possible to have a matrix organization within a matrix organization (for example, the matrixed engineering department). In addition, the extra functional unit (that is,

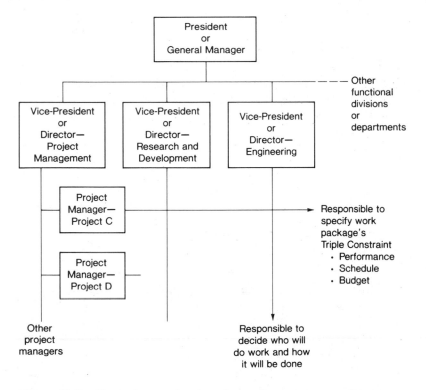

Figure 12-3.   Typical organization chart of a matrix organization.

project management) can proliferate bureaucratic tendencies, and the balance of power between project management and functional units can exacerbate conflicts.

OTHER ORGANIZATIONAL FORMS

Quasi Matrix

The quasi matrix (Figure 12-4) is a compromise way to obtain the benefits of the matrix form in a functional organization otherwise too small to afford it. In this compromise organization, when someone is designated project manager, he or she remains part of the functional group for project work done in that group (which is the same as in the functional organization). However, for project management work, he or she reports directly to top management.

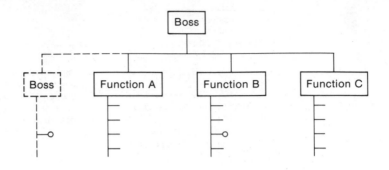

Figure 12-4.　Quasimatrix organization.

Source: "Basic Management Skills for Engineers and Scientists." Lecture notes by Milton D. Rosenau, Jr. Copyright © 1982 by the University of Southern California. Reprinted by permission.

Thus, the project focus has top management's support to cut across functional group boundaries. In some cases, the quasi matrix will have a manager of projects (just as in the matrix organization), rather than the boss, to whom the project managers report for project (as distinct from functional) matters.

Venture

The venture organization, common in several very large, commercially oriented companies, is especially appropriate for projects aimed at new product development. Basically, the goal is to set up a tiny functional organization within a giant corporation, thus achieving the advantages of compact size, flexibility, and the entrepreneurial spirit of the small company within and supported by the financial, physical, and human resources of the larger company. Where such a management organization exists, it is common to team up an engineer with a marketing person and a manufacturing person in the earliest phases of new product development. As the effort moves forward, the venture organization grows, ultimately becoming a functionally organized division within the parent company. Two practitioners of this managerial form are 3M and Dupont.

Task Force

Organizations frequently use a task force to cope with an unexpected project. Hence, this response is most commonly utilized by a functional organization because the other organizational forms are already able to deal with projects. A task force may be thought of as a rarely used single project organization within the functional organization. It can be formed quickly, usually by a very senior officer.

A task force may be helpful in coping with an unexpected project.

Although the people selected to serve on the task force may be highly motivated by their selection, they frequently are not relieved of their usual duties and thus may not have sufficient time for the task force. If they are relieved of their normal duties, they may be anxious about their assignment when the task force has completed its job.

THE INFORMAL ORGANIZATION

Regardless of the formal organizational structure, there is always an informal organization (see Figure 12-5). Where friendships and common interests are present, these create channels through which information flows easily and cooperation is encouraged. Conversely, rivalries and animosities can inhibit cooperation.

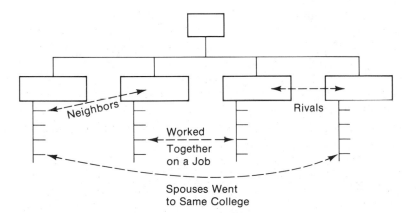

Figure 12-5.   The informal organization.

Source: "Basic Management Skills for Engineers and Scientists." Lecture notes by Milton D. Rosenau, Jr. Copyright © 1982 by the University of Southern California. Reprinted by permission.

TYPICAL PROBLEMS

> Each organizational form has its advantages and disadvantages. The only real problem occurs when a project manager believes that a different organizational form will solve all the organizational problems he or she is experiencing. In actual fact, no organizational form is perfect for all situations or for all time. Thus, reorganization is regularly required in any organization doing project work to keep the organization as useful and productive as possible.

MATERIALS STUDY PROJECT

> The organization chart for SUPROMAC is illustrated in Figure 12-6. This is a functional organization. Therefore, Mel Chase will probably have problems getting support from other groups and sections unless a quasi matrix is established.

HIGHLIGHTS *Three common organizational forms for project management are functional, project, and matrix.*

Three other forms that may be used are quasi matrix, venture, and task force.

The informal organization is always present and affects how well the formal organization works.

FURTHER READING

T. M. Adams. "Matrix Management—Panacea or Pandemonium?" *IEEE Engineering Management Review*, vol. 8, no. 1

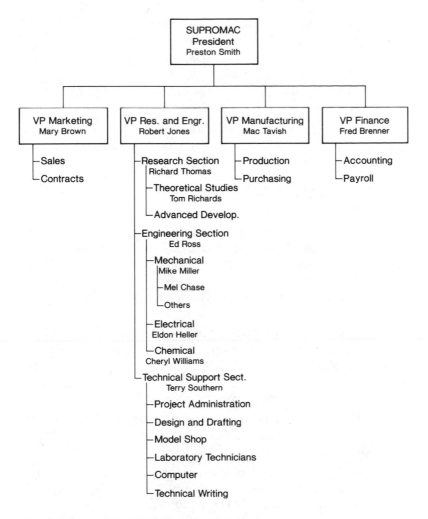

Figure 12-6. SUPROMAC organization chart.

(March 1980), pp. 55–64. (Reprinted from *Personnel Psychology*, vol. 30, no. 1 [Spring 1977], pp. 55–64.)

This breezy article compares functional, project, and matrix organization forms in terms of very practical issues.

R. D. Archibald. *Managing High-Technology Programs and Projects.* New York: Wiley-Interscience, 1976.

Chapter 5, section 4, is a thorough discussion of organizational options with many examples.

S. M. Davis and P. R. Lawrence. "Problems of Matrix Organizations." *Harvard Business Review,* vol. 56, no. 3 (May-June 1978), pp. 131–142.

The article summarizes the main points in their book, providing a thorough treatment of the matrix form.

F. L. Harrison. *Advanced Project Management.* Aldershot, Hants, England: Gower, 1981.
Chapter 12, pages 275–312, discusses human behavior in the project setting.

W. Jerkovsky. "Functional Management in Matrix Organizations." *IEEE Transactions on Engineering Management,* vol. EM-30, no. 2 (May 1983), pp. 89–97.
Although academic in tone, this article touches upon some of the problems functional managers have in a matrix organization.

H. E. Pywell. "Engineering Management in a Multiple- (Second or Third Level) Matrix Organization." *IEEE Transactions on Engineering Management,* vol. EM-26, no. 3 (August 1979), pp. 51–55.
This is a good discussion of the matrix's value in larger projects.

W. E. Souder. "Project Management: Past, Present, and Future—An Editorial Summary." *IEEE Transactions on Engineering Management,* vol. EM-26, no. 3 (August 1979), pp. 49–50.
This is an overview of matrix pros and cons.

M. A. Verespej. "Mission Extraordinary? Call for a Task Force." *Industry Week* (October 19, 1981), p. 67.
This is a brief article on ad hoc project management.

W. C. Wall, Jr. "Integrated Management in Matrix Organization." *IEEE Transactions on Engineering Management,* vol. EM-31, no. 1 (February 1984), pp. 30–36.
This is a conceptual overview of how a project interacts with an organization.

R. Youker. "Organization Alternatives for Project Managers." *Management Review* (November 1977), pp. 46–53.
This is a succinct and graphic summary of organizational options.

13

Organizing the Engineering Project Team

The project team consists of those who work on the project and report administratively to the engineering project manager. This is in distinction to the support team (people who work on the project but do not report administratively to the engineering project manager), which I discuss in Chapter 14. In this chapter, I first review sources of project personnel and consider the frequent necessity to compromise by using whoever is available. Then I deal with how much control an engineering project manager can exercise over project personnel and provide some practical tools to help him or her gain effective control. The last section discusses the use of task assignments both as a means to assign the work packages and to obtain commitments from personnel to carry out the work.

DEGREE OF ASSOCIATION WITH THE PROJECT

Table 13-1 shows eight categories of personnel assignment to projects. They result from all possible combinations of three factors: (1) whether personnel report directly to the project manager or are administratively assigned to someone else, (2) whether they work full-time or only part-time each day (or week or year) on the project, and (3) whether they work on the project from its inception to completion or for only some portion of the project.

TABLE 13-1. The project team and the support team.

Nature of Reporting Relationship / Duration of Project Assignment	Reports to Project Manager		Works on Projects but Reports to Another Manager	
	Works Only on Project	Also Has Other Assignment(s)	Works Only on Project	Also Has Other Assignment(s)
From Start to Finish	P	P	S	S
Only a Portion of Project's Duration	P	P	S	S

P = Project Team
S = Support Team

Project Team

The project team is composed of the people who report administratively to the project manager (the four cells so designated on the left side of Table 13-1). I consider this the project team because the project manager can assign work packages to these people rather than having to negotiate with other managers to obtain commitments for their work.

Team Matrix

The amount of project labor obtained from each category depends on the project contractor's organizational form (that is, functional, project, or matrix) and project size. In a matrix organization, no one may be assigned to work for the project manager; the entire labor pool may be drawn from the support team. In a pure project organization, the vast majority of project labor, perhaps all of it, may be assigned to the project manager. This is especially likely for a large project of long duration. As a practical matter, small projects are not likely to have their own personnel, regardless of organizational form. The majority of those assigned to a project from start to finish is either managerial or administrative because most other skills are required for only some portion of the project.

A key point that emerges from consideration of Table 13-1 is that the project manager must provide eight different kinds of management attention to people working on the project. People who work on the project for only a portion of its duration must be managed to be ready when needed; then orientation to the project must be provided; finally, the engineering project manager must recognize that they may be frustrated or lack a sense of accomplishment at leaving the project prior to completion. People who have other assignments must be persuaded that the work on the project deserves their attention each day (or week). Because they may have a lower stake in the project, they often require better or more forceful leadership. People on the support team must be managed through other managers, which can lead to distorted work direction. Issues of priorities, performance standards, and loyalty often require the engineering project manager's attention.

There are eight ways to assign personnel to a project; so the project manager must manage differently, as required by each way.

SOURCES OF PERSONNEL

There are many sources of people, including the proposal team, other people already employed by the organization, and people from outside the organization (hired personnel, contract personnel, consultants, and subcontractors).

The Proposal Team

By far the best source of project personnel are those who worked on the proposal. They are familiar with the subject matter of the proposal and perhaps to some extent with the customer's specific problems. Being already "up to speed," they require minimal indoctrination. They will, for example, presumably understand the meaning of potentially ambiguous words (such as "gain").

The proposal team is the best source of project personnel.

Other Organizational Employees

Other employees of the organization are the second best source of personnel. These people are at least familiar with company policies and procedures; they know where the library, the model shop, and such are located. Although they may not be familiar with the specific subject matter of the proposal, they are at least familiar with how the company does business and know its strengths and weaknesses. They know whom to call for help and where to go to get something done. In fact, they have probably worked on similar projects in the past.

The project manager may know their strengths and weaknesses and thus be able to assign them to appropriate work packages. The project manager will not ask them to do more than their capabilities permit or give them a work package so trivial as to be demeaning or demotivating.

People from Outside the Company

There is a variety of outside sources for personnel (see Figure 13-1). Consultants, contract ("body shop") personnel, and subcontractors can be obtained quickly, but it often takes months to hire a new person. The newly hired person, who can work for and be fully under the administrative control of the project manager, may be thought of as part of the project team in the sense I am using that term. To hire a person, a project manager has to have a personnel requisition approved, typically must advertise the position, interview several people, make one or more offers to get an acceptance, and wait for the person to relocate (if required) before coming to work. Then there is an indoctrination period while the person becomes familiar with company practices. A project of short duration rarely can afford the time to hire personnel and thus depends on the support team.

Permanently	Hiring
	Transfer
Temporarily	Consultant Job Shop Subcontract

Figure 13-1.   Filling personnel needs.

Source: "Basic Management Skills for Engineers and Scientists." Lecture notes by Milton D. Rosenau, Jr. Copyright © 1982 by the University of Southern California. Reprinted by permission.

COMPROMISE

It is indeed rare that an engineering project manager can staff the project entirely with personnel who (1) already work for him or her, (2) worked on the proposal, and (3) represent exactly the right distribution of skills to carry out the project. Usually, the project manager must staff the project from whoever is currently available either full- or part-time. Many of these people will not completely meet the requirements. It is often a case of fitting square pegs in round holes.

Staffing compromises are usually necessary.

Qualifications

The newly appointed engineering project manager confronted with the urgent need to staff a project team is often victimized by other managers in the company who offer their "cats and dogs." These people may be marginally employed; so company management may pressure the project manager to accept them into his or her group.

This is a very tricky situation. There is pressure from above to accept the people, and there is another manager offering them as freely available. But, if these people are known to be marginal workers, it is probably better to terminate their employment than to shift them from one project to another, burdening these projects and retaining marginal workers for long periods of time. Nevertheless, it is common for a newly appointed engineering project manager to be offered all kinds of personnel for transfer. On a short duration project, it may be better to accept these workers, unless they are clearly unqualified, than to recruit better qualified assistance.

Motivation

Some projects that offer high pay (such as the trans-Alaska pipeline or projects requiring a great deal of overtime or shift premium work) frequently attract workers whose primary motivation is money. The engineering project manager may be besieged by candidates who wish to go to work on his or her project. Their motivation, however, may not be best for the project. So confronted, the manager should seek to staff the project team with a few high quality people and confine the money seekers to support team roles, where they are someone else's problem.

Conversely, a project with high scientific content or one of national importance (such as the Apollo project) often attracts highly dedicated, altruistic people. A common correlate of this altruism is a lack of practicality, which the project manager must watch for and temper.

Some project personnel may be poorly motivated and some may be unrealistic.

Some projects have an unsavory reputation (fairly or unfairly earned) that makes it very difficult to recruit personnel. They often require portions of the work be performed at an unattractive or remote location. To overcome this drawback, various inducements may be required.

Recruiting Qualified Help

Some compromise is clearly required in staffing the project team, but there may be some skill requirements that cannot be compromised.

Most engineering project managers prefer to have people on the project team because it seems to improve project control. People on the project team cannot be given other distracting work assignments unless the project manager approves it, but people on the support team may be given other work that detracts from their ability to honor support commitments.

Thus, in many situations where you want the best qualified help for your engineering project, you must locate a person with the required skill and try to obtain a transfer (perhaps temporary). In some cases, the transfer request will be refused. If your organization has a matrix form, of course, transfer is not possible; at most, you can obtain a firm commitment to assign a needed, uniquely skilled person only as long as the project requires him or her.

CONTROL

Supervision

In talking about the project team, I am talking about people who work for the project manager. They may not work full-time on the project, either for its entire duration or full-time within any given workday or workweek. Nevertheless, they are under the project manager's direct supervision, unless there are intermediate levels of supervision. Some of these people may have been transferred from other managers.

Projects go through different phases, which implies that

personnel must be changed. For instance, the creative engineering design person, so valuable in the early phases of
system design, is not needed when the project is moving
toward completion and the team is trying to finish what has
been designed rather than figure out additional clever ways
to design it.

Thus, some project team people may have to be reassigned
during the project. An administrator or junior project manager assigned to work under the project manager may be
needed the entire time, but other personnel may need new
assignments. They will either go to work on another project
full-time after completing their work or work on two or three
projects part-time. It is therefore important that the project
manager exercises control over the timing of these assignments so as to have people with the right skills available
when required and have other assignments for them when
they are not required. This is one reason a resource allocation
analysis is desirable.

*People must join
and leave the
project as needed.*

Proximity

One of the project manager's most powerful tools for improving control of project personnel is to locate everyone in a
common area. This aids communication, and, where there is
increased communication, there is increased understanding.
This also improves the likelihood that everyone on the team
understands the Triple Constraint.

Problems

Most project managers would rather staff the project with
many project team members and fewer support team members. However, this staffing forces the project manager to
contend with many personnel problems, such as people quit-
ting (either the company or organizational unit) to work elsewhere, sickness, higher management reassignments, lack of
interest in the project, or other conflicting assignments.

*Most projects
cannot be fully
staffed with
project team
members.*

TASK ASSIGNMENTS

I have previously emphasized use of the work breakdown
structure and network diagram to divide projects into small
pieces of work. Each of these pieces, or tasks, has a corres-

ponding cost estimate. In the ideal world, the person responsible for each task has prepared both the schedule and the budget estimate. This person should also have played a significant role in defining the exact Triple Constraint of his or her small work package. In any case, the project manager must assign tasks to many different people. As these tasks are assigned, some give and take in the exact scope may be accepted, but whatever is finally agreed upon must be committed to paper. That is, there should be minicontracts between the project manager and the people responsible for tasks. The minicontract defines the Triple Constraint of the task.

All work assignments should be written.

The project team member who now has his or her task assignment should provide the project manager with a detailed plan of how that task will be performed and periodically review progress against the plan. To the extent that the task performer has played a major role in creating and initiating the task assignment, he or she is likely to be highly motivated to carry it out. Conversely, if the task was assigned without negotiation, the person may have a low sense of involvement and be largely demotivated by the assignment.

TYPICAL PROBLEMS

> The usual problem is what to do with marginal personnel. This is one reason you should have inserted schedule and budget contingency—because sometimes you have personnel who must be used in an area outside their competency, which renders them temporarily marginal. In the case of truly marginal personnel, you can simply refuse to accept them on the project team.

MATERIALS STUDY PROJECT

> Imagine you are Mel Chase, SUPROMAC project manager for the materials study project I have been using
> *continued*

as an example. Because Tom Richards is the only person in SUPROMAC capable of doing the theoretical study (tasks G, H, and J), you must have him conduct these tasks. You have two options: (1) Have the director of research transfer Tom Richards to your project team for the duration of the project. (2) Have Tom Richards perform the work without reassignment, as part of the support team. Thus, you would first try to persuade Richard Thomas to permit the transfer and then try to persuade Tom Richards to accept it. You could do this by stressing the importance of the project, the importance of Tom Richards's contribution to it, the reduced demand such a transfer would make on his supervisory time, and the temporary nature of the transfer. You might also stress how the experience Tom Richards would gain would benefit both him and the research section when the temporary transfer ended.

Nevertheless, such a transfer could well be undesirable for the project if you do not need Tom Richards 100 percent of each day. What would he do the rest of the time? It can also be undesirable if he is a difficult person to supervise. But if these are not obstacles, you should try to persuade Richard Thomas to approve the transfer.

Or imagine that SUPROMAC's proposal, and the resulting contract with NERESCO, designate Molly Cook (a noted mechanical expert) to work on the job, but she has just begun a long hospitalization. Your options include arranging for her to work while hospitalized, even though this may be inefficient and therefore more costly; trying to substitute a consultant or contract personnel; or trying to substitute less qualified SUPROMAC personnel.

HIGHLIGHTS

The project team is people who work on the project and report administratively to the project manager.

Sources of project personnel include the proposal team, others employed by the organization, and people from outside the organization.

Compromise is required in forming project teams.

People must join and leave the team as required during the project.

Having team members in close proximity improves the project manager's control.

FURTHER READING

C. L. Buck. "Managing the Most Valuable Resource: People." *Project Management Quarterly,* vol. 8, no. 2 (June 1977), pp. 41–44.

   This is a useful, brief discussion of building project teamwork.

C. Cammann and D. A. Nadler. "Fit Control Systems to Your Managerial Style." *Harvard Business Review,* vol. 54, no. 1 (January-February 1976), pp. 65–72.

   This article contains excellent suggestions for improving control of personnel.

H. J. Thamhain and D. L. Wilemon. "Leadership, Conflict, and Program Management Effectiveness." *Sloan Management Review,* vol. 19, no. 1 (Fall 1977), pp. 69–89.

   This article has research data on effective management techniques for project managers.

Organizing the Support Team

The support team is the people who work on the project either full-time or part-time for a part or all of the project but do not report administratively to the project manager. This chapter discusses how to obtain their involvement and commitment and how their efforts can and must be coordinated with the project team. Then I consider interaction between the project team and support groups and subcontractors.

INVOLVEMENT AND COMMITMENT

As with the project team, the best way to develop a sense of involvement and obtain a commitment from the support team is to have had its members participate in the proposal. Participation also builds a team spirit that continues beyond the project. Failing this, their involvement in planning their own work and committing those plans to writing should also elicit involvement and commitment—remember the Golden Rule.

Involve support team members in the proposal phase or as early as possible.

Early Support Group Involvement

Project managers and the project team often ignore support requirements, which other groups must provide, until it is too late. Unless support personnel understand that their services may be required, they cannot anticipate the extent to which they will be needed. Consequently, the support a project demands may not be available when needed. When support is sought tardily, support groups feel left out, and it may be difficult to obtain their commitment.

This kind of situation may arise because the project team has some degree of parochialism or is not aware what support is readily available. The project team may not understand the potential roles others can play or may assume it knows better than the support groups what kind of effort will be required. This latter situation frequently arises because the project team feels that a support group will "gold plate" the amount of work they propose to do, exceeding project budgets.

Have support team members estimate time and cost of their tasks.

As stated earlier, these problems can best be mitigated by involving support groups in the proposal phase. If this cannot be done, involve them as early as possible in the project work. Give them an opportunity to participate in planning their task and employing their best thinking and expertise.

The same applies to the time and cost estimates. The support group should make time and cost estimates for their task, and the project group should approve them. These estimates may require a negotiated revision to adjust other project tasks to accommodate support group plans if they differ from the project team's first estimate. This is a common occurrence. Support groups sometimes must perform their role at a pace dictated by other, higher priority commitments, thus scheduling your project support differently than you had planned. Sometimes the support group sees a completely different way to undertake its role, often to the project's advantage. Or the support group's experts may convince the project team that their role must be broader than originally conceived. For all these reasons, involve support groups as early as possible.

Written Commitments

Put all agreements in writing.

Obtain meaningful commitments from support groups within your organization just as you do from the outside subcontractors, namely, a written agreement. (This is also what you should do with project team member commitments; the only difference is what actions you can take to settle disputes that may arise.) There must be a Triple Constraint and signatures by both parties. Such agreements (inside the organization) lack legal standing and enforcement provisions, but if the support group manager must sign a written agreement, he or she will be motivated to make his or her group live up to its commitment.

Support Team Advantages

As I said in the previous chapter, most project managers seem to prefer to staff their project entirely (or mostly) with project team members. However, a project manager (especially in a matrix organization) might prefer to have a large support team rather than a large project team, for the following reasons:

1. The project manager does not have to worry about the support team after the project ends.

2. In the case of subcontractors, the support agreement is embodied in a legally binding instrument, namely, the subcontract or purchase agreement.

3. The project manager has the whole world in which to find specialists or experts with the required skill.

COORDINATION

Once the support groups have been identified and their work has been planned properly and phased in with that of the project team, there is a continuing need to coordinate project team work. This is best done with network diagrams (Figures 14-1 and 14-2). In both figures, support group work has been segregated from the main part of the network. There are many other ways to do this, for instance, using distinctive line patterns for each support category. Where color copying machines are available, a color code may be used advantageously.

A time-based CPM diagram aids coordination.

Change

Consider the network diagram of Figure 14-2. Imagine that you discover you will be late on task D. Should you inform your subcontractor (on task B) of this lateness? It depends. In general, it is probably best to advise subcontractors of your true need date. If you do this, you make it easier for them, and their costs to you will be lower (at least in the long run). But if subcontractors have a history of lateness, it is probably best (1) to have originally allowed time contingency for their work and (2) not to let them know of any delay you have experienced.

Communication and coordination should be primarily in

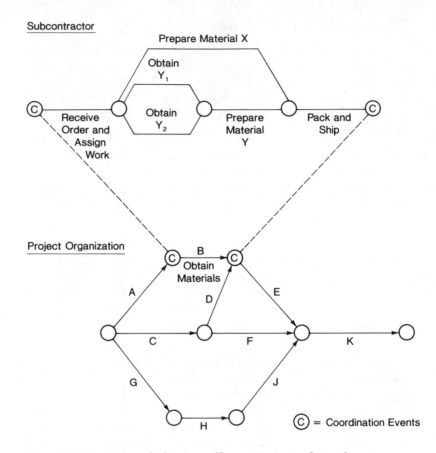

Figure 14-1. Network diagram illustrating use of coordination events.

writing. Change should be accomplished by oral communication, over the telephone and/or at meetings involving as many people as required. But the change must then be embodied in the plan revisions.

Revision

Write and distribute plan revisions.

Once committed to paper, plans must be disseminated to and understood by all involved personnel. Plans must also be maintained in a current status. If any out-of-date project plans are allowed to remain in circulation, the credibility of all project plans will become suspect. Therefore, everyone who had the original plan must receive revisions. This can be facilitated by keeping an accurate distribution list.

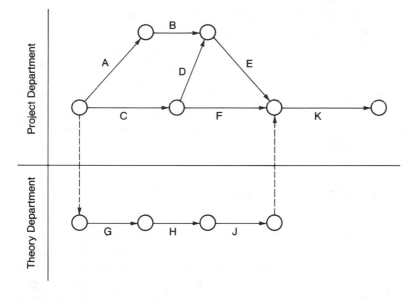

Figure 14-2.   Network diagram illustrating use of spatial segregation of a support group's activities.

INTERACTION WITH SUPPORT GROUPS

Project team and support group interaction can be difficult. All too often the support group is brought in too late, a situation that reminds me of a story. A commuter comes dashing onto the train platform just as the morning train into the city pulls out. A bystander, observing that the commuter has just missed the train, comments, "Gee, that's too bad. If you'd simply run a little bit faster, you'd have caught the train." The commuter knows, of course, that it is not a matter of having run a little bit faster but rather of having started a little bit sooner.

Project Actions

The tardy commuter's situation classically applies where the purchasing department is involved. Purchased materials arrive later than required; the project is delayed; and project people blame either the subcontractor for delivering late or the purchasing department for failing to place the order early enough. In fact, the blame lies with the people who did not requisition the purchase sufficiently early so the goods would be delivered on time. They need not run faster; they should have started earlier.

The experienced engineering project manager copes with this problem in two ways. First, he or she makes certain that the network diagram schedule allows enough time for the support groups to perform optimally. Second, the project manager makes certain that all personnel know when task activities must be completed and holds the task managers accountable for meeting the schedule.

Support Team Viewpoint

These issues can be looked at in a different way, namely, from the point of view of the support groups. They are composed of professionals, in the previous example, purchasing professionals who wish to obtain the best quality of required goods at the lowest possible price within the other constraints project personnel impose. They need time to perform their function in a professionally competent way. In the case of government contracts, there are even laws and regulations that require three competitive bids.

Support groups labor under many constraints.

The same is true, of course, of any support group—technical writers, computer programmers, designers, draftpersons, or model shop personnel. Everyone wants to do a good job and wants sufficient time in which to do it. But departments have a work load imposed on them by others. They are trying to respond to many projects bringing work to them at random times and in variable amounts. Thus, support groups typically have some backlog they must work through before they can get to new requests. If they did not have this backlog, if they were sitting there idly waiting for the work to arrive, they would not be utilizing a vital organization resource, their own time, in the most effective way.

SUBCONTRACTORS

Subcontractors are basically no different than you. They have a contract from a customer, in this case, you or your engineering project. They want to be responsive to you, but they have the same kinds of problems you do: Personnel and resources frequently are not instantly available or perfectly suitable; they need time to plan their work; they have to interpret the Triple Constraint in the correct way, and so on.

Just as a contract controls your relations with your customer, so subcontractors define their relationship to your project by the contract your company's purchasing department issues to them. Should a change be required, it is certainly all right to tell them about it. But the change becomes effective and meaningful only when it is converted into a contract change.

Another point to consider when working with subcontractors is that your request for a proposal can require that periodic reviews be included in your contract. This is desirable, as it would be if your customer required periodic reviews of your work. You are trying to see how their work is progressing, to understand if changes will be called for as a result of what they are doing or problems they are encountering, in short, to stay abreast of their work.

In many cases, you can do other things to help your subcontractors perform effectively. For instance, you may have one or more lengthy meetings with their project manager to be certain he or she knows what is important to you. You can be willing to compromise noncritical items (with suitable contract changes, if appropriate). You can understand and review their schedule to try to offer constructive suggestions. You can check analyses and witness engineering tests, provided you have made suitable arrangements.

But you must draw a fine line between giving new directions and simply keeping abreast of what they are doing. Remember, the contract dictates the work. The progress reviews or monitoring activities are not a substitute for their management of their work; rather, these are solely to find out if it is being done. You can not provide daily, weekly, or monthly changes in their direction and expect them to be successful.

Your support agreement is a written contract.

TYPICAL PROBLEMS

Working with the support team probably causes the greatest difficulty, especially for new, inexperienced engineering project managers. The root of this problem is being dependent on nonsubordinates. Two other problems are closely related. First, to negotiate support agreements takes a lot of time, usually at the very busy
continued

project inception period. So it is done reluctantly or poorly or even omitted. In the latter situation, the project manager uses his or her own judgment of what the support group will do. Second, even when the support agreements have been intelligently negotiated, later events frequently require that changes be made. Again, this is time-consuming and must be anticipated.

MATERIALS STUDY PROJECT

Consider the materials study project I have been using as an example. Let us make two further assumptions about it. First, again assume that task B ("obtain materials") will require a subcontract to a material supply company. This procurement will be governed by a written contract or purchase order. Such a written instrument is valid only when it is accepted by the material supply company, usually acknowledged by the return of a signed copy. There can be no agreement unless both parties are satisfied by the contemplated Triple Constraint.

Second, assume that Tom Richards remains assigned to Richard Thomas but will be responsible for the theoretical tasks (tasks G, H, and J). All too often this kind of agreement is only verbal. But why should a $77,740 theoretical effort (see Figure 9-15) be loosely defined when a much smaller material supply effort is governed by a legally binding agreement? Support group work is commonly authorized informally, and most of the time the results are a disappointment, if not a shock, to the project team.

HIGHLIGHTS   *Support teams do not work for the project manager in an administrative sense, but their participation and contributions are vital.*

Support groups should be involved in projects as early as possible and allowed to plan their task.

Although they lack legal status, written agreements are an excellent way to obtain commitments from support teams.

Coordination, a continuing need, is best provided by network diagrams.

Every plan revision must be written and distributed to all concerned personnel.

FURTHER READING

R. D. Archibald. *Managing High-Technology Programs and Projects.* New York: Wiley-Interscience, 1976.
  Chapter 8 reviews many of the written devices the project manager can use to control the support team.

A. J. Melcher and T. A. Kayser. "Leadership Without Formal Authority—The Project Department." *California Management Review,* vol. 13, no. 2 (Winter 1970), pp. 57–64.
  This is a case history, with some useful ideas on how to obtain cooperation from support groups.

15

The Role of the Engineering Project Manager

Although the project manager is clearly involved in all phases of the project and is ultimately responsible for satisfying the Triple Constraint, his or her interaction with the project and support teams is a key to the leading (or "people management") phase. This chapter first examines the overriding importance of the project manager's ability to influence other team members. His or her leadership ability depends on motivational skills rather than on authority, regardless of how much hierarchical supervisory authority he or she has over project team members. Because projects are one-time undertakings, engineering project managers must often develop creative solutions to project problems. I suggest ways to stimulate creativity.

WHAT AN ENGINEERING PROJECT MANAGER DOES

Influence Rather Than Authority

As the three previous chapters note, many people working on an engineering project do not report directly to the project manager, and he or she does not even have complete control over those who do. In the first place, people are free to change jobs in our society. If given a command they do not like, some workers will simply quit. Or they may transfer to another division of the organization. Second, modern motivational theory indicates that issuing commands is a poor means to encourage people to perform well on a job. McGregor's Theory X/Theory Y is but one manifestation of the thinking that underlies current managerial practice, which usually substitutes persuasion and participation for command.

The project manager lacks control.

Nevertheless, commands are still a way of life, to a greater or lesser extent, depending upon the specific organization and situation. If stated brutally or insensitively, they demotivate and create resentment. If stated politely and reasonably (which is difficult to accomplish), commands may be effective.

Given these limits to hierarchical authority, engineering project managers must operate by winning the respect of project and support team members. This accomplished, they will find their wishes are carried out voluntarily and frequently with enthusiasm.

A study by Thamhain and Wilemon identifies nine influence bases available to project managers.

There are nine ways to have influence.

1. Authority—the legitimate hierarchical right to issue orders

2. Assignment—the project manager's perceived ability to influence a worker's later work assignments

3. Budget—the project manager's perceived ability to authorize others' use of discretionary funds

4. Promotion—the project manager's perceived ability to improve a worker's position

5. Money—the project manager's perceived ability to increase a worker's monetary remuneration

6. Penalty—the project manager's perceived ability to dispense or cause punishment

7. Work challenge—an intrinsic motivational factor capitalizing on a worker's enjoyment of doing a particular task

8. Expertise—special knowledge the project manager possesses and others deem important

9. Friendship—friendly personal relationships between the project manager and others

The first clearly depends on higher management's decision to invest the engineering project manager with power, regardless of power's intrinsic utility. The next five may or may not be truly inherent in the project manager's position; others' perceptions are most important in establishing their utility to the engineering project manager. The seventh is an available tool anyone may use to influence others. The project manager must earn the last two. Projects are more likely to fail when the project manager relies on authority, money, or penalty to influence people; success is correlated with the use of work challenge and expertise to influence people.

There will be occasions when the project manager must negotiate with team members. A typical approach is to explain the rationale of the effort and to involve the people in planning the detailed work packages. Given this need to influence, an effective project manager must be a superb communicator. He or she must have verbal and written fluency and be persuasive to be effective; the next chapter contains some practical tips on how to improve your communication skills.

Effective Managerial Behavior

As I said in Chapter 1, the project manager must work with people not of his or her own choosing, many of whom have different skills and interests. Furthermore, the project manager is a manager, not a doer. If the project manager is writing detailed test specifications for a structural design project, designing a circuit for a new product project, or operating the pilot plant for a new process, who is planning the work of others? Who is deciding what approach to take to the support group manager so as to obtain the services of the most senior and best qualified person? And who is trying to devise a contingency plan in case the system test does not produce desirable results? The project manager must spend his or her time working with people and planning their work so nothing is overlooked and contingency plans are ready if needed.

A manager must plan and manage.

On a very small project, the project manager's participation is also required as a worker, not merely as a manager. If not physically, then at least mentally, a project manager in this situation should have two hats, one labeled "project manager" and the other labeled "worker." The project manager must realize which function he or she is performing at any given moment and wear the appropriate hat.

Qualifications

Generally speaking, one becomes a project manager because one has been an excellent mechanical, electrical, optical, chemical, or industrial engineer rather than because one has been trained or demonstrated competency as a project manager. But a virtuoso technical performance is not a sufficient qualification for managing the efforts of the project and support teams. In fact, one's demonstrated technical or professional skills are frequently problem-solving or technical skills that do not involve an ability to interact with others. But project managers, in common with other managers, need people skills rather than technical skills (see Figure 15-1). Developing people skills can be extremely difficult for many technically trained people who become project managers. Physical systems tend to behave in repeatable and predictable ways; people do not.

The engineering project manager must deal with many intangibles.

Working with People

Elias Porter has shown that we behave differently when everything is going well and when we face opposition or conflict. People also differ from each other. Some have

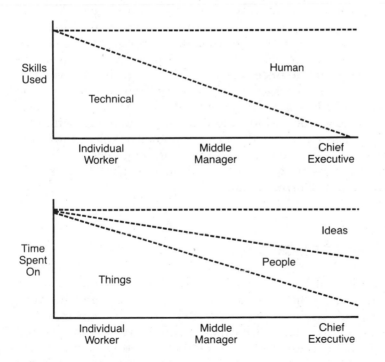

Figure 15-1. Management.

Source: "Basic Management Skills for Engineers and Scientists." Lecture notes by Milton D. Rosenau, Jr. Copyright © 1982 by the University of Southern California. Reprinted by permission.

an altruistic-nurturing orientation; other people have an assertive-directing orientation; and others have an analytic-autonomizing orientation. Although altruistic-nurturing oriented people are usually trusting (a strength), people of another orientation may see them as gullible (a weakness). Similarly, the assertive-directing person's self-confidence (a strength) can be seen as arrogance (a weakness), and the analytic-autonomizing person's caution (a strength) may appear to be suspicion (a weakness). A person with a balance of these orientations, who is flexible, may be seen as inconsistent.

The project manager should be chosen because of an interest and skill in human relations.

To compound this problem, the same thing said to the same person at two different times can produce different reactions. This lack of predictability can be a major pitfall for many prospective project managers. Project managers must deal with both technical and emotional issues. If not already fluent with these human relations skills, they should take a course in behavioral psychology.

Managers manage and workers perform the tasks.

The project manager sets objectives and establishes plans, organizes, staffs, sets up controls, issues directives, spends time working with widely varied people, and generally sees the project is completed in a satisfactory way, on time, and within budget. The project manager does not do the work of others on the project. A project manager who is an excellent electronic engineer may find it frustrating to watch a junior engineer carry out the circuit design activities on the project. The junior engineer will take longer, make mistakes, and not do as good a job as could a project manager with that technical skill. But if the project manager starts to do the circuit design, it demotivates the junior engineer and lessens the manager's time to function in the most vital role of all, namely, that of project manager.

THEORIES OF MOTIVATION

Regardless of hierarchical authority, anyone can encourage or stimulate others to contribute to a project and improve their productivity. For engineering project managers, many of whom lack direct authority over all the resources required for project success, an understanding of motivation is essential. A major element of the project manager role is to avoid demotivating others. There are many theories of motivation, the two most important of which I briefly review.

Hierarchy of Needs

Abraham Maslow's theory of the hierarchy of needs (see Figure 15-2) holds (and human experience validates) that motivation is not external but rather arises within the worker if managerial actions are not inappropriate. The worker is motivated to achieve a specific goal because of an inherent internal need. Each person has five levels of needs. The first of these is physiological or body needs (eat, sleep, have shelter, and so forth). A hungry person will have a goal to eat and will engage in the goal-directed activity of buying and preparing food to satisfy the hunger need. (In fact, if we try to prevent this person from satisfying the need, by denying him or her money to pay for the food, for instance, the person may engage in such antisocial behavior as robbery to satisfy the need.)

People are motivated to fill their unsatisfied needs.

Once this need is filled, continuing to offer more food, sleep, or shelter has no motivational value. Higher level needs now come into play. Second-level needs are safety and security; third-level needs are social; and fourth-level needs are esteem or ego. The fifth-level needs are for self-actualization or fulfillment. To take advantage of Maslow's findings, an engineering project manager would have to understand the

People have five levels of needs.

A Satisfied Need is No Longer a Motivator

Figure 15-2. Motivation—Abraham Maslow's hierarchy of needs.

Source: "Basic Management Skills for Engineers and Scientists." Lecture notes by Milton D. Rosenau, Jr. Copyright © 1982 by the University of Southern California. Reprinted by permission.

levels of need an individual has already satisfied. Then he or she could offer satisfiers for unmet needs as an encouragement. Americans generally have unfilled needs on levels three, four, or five; therefore, the project manager would offer satisfiers aimed at these levels.

If the project manager does not offer satisfiers, workers will find them outside the work environment (see Table 15-1). Note that money is not a motivator, although it may play a role. A person feels good when he or she gets a salary increase because it provides recognition (filling an ego need). However, if a person can fill social needs only by joining a country club, he or she needs more money than the person who can satisfy these social needs with informal activities.

Motivational Factors

Frederick Herzberg has done an excellent study on work and motivation, examining specific factors that motivate workers. He found many things done (company policies, supervision, work conditions, and salary, for instance) are not motivational at all. He called these "hygiene factors." The absence of hygiene factors is demotivating, but their presence is not motivating.

TABLE 15-1.   Satisfiers.

Level	Job Related	Other
5	Setting own goals	Hobbies or volunteer work
4	Winning company award	Election to civic board
3	Staff meetings or committee work	Club membership
2	Insurance packages	Live in safe area
1	Enough salary to purchase necessities and reasonable work hours	Availability of adequate stores

Source: "Basic Management Skills for Engineers and Scientists."
Lecture notes by Milton D. Rosenau, Jr. Copyright © 1982 by the University of Southern California. Reprinted by permission.

Instead, there have to be motivational factors present to foster high productivity. The key motivational factors are achievement, recognition, the work itself, responsibility, advancement, and growth. Achievement and recognition are short term, and the others are long term in their impact. Thus, it is important to give workers recognition frequently (but not routinely) for significant accomplishments.

Achievement and recognition are the most powerful motivators.

IMPLICATIONS OF MOTIVATIONAL THEORY

How can you take advantage of these widely accepted theories of how people become motivated? There are many techniques, three of which I briefly review.

Theory X/Theory Y

I have already mentioned Douglas McGregor's work in identifying two managerial styles, Theory X and Theory Y. Theory X is the authoritarian style, in which top management makes decisions and coerces workers to comply. Theory Y is the participative style, built upon the findings that people both enjoy working and want to work. Theory X assumes external control of the workers; Theory Y assumes useful controls are within individuals and managers can draw upon workers' self-direction.

In working with the support team, over which he or she lacks control, the engineering project manager should try to adopt the Theory Y style (and would probably be well advised to use it with the project team as well). Authoritarian behavior is not usually appropriate. Managerial behavior can be based on the Theory Y assumption that others want to do a good job.

Participative decision making is best.

To put it another way, consider how you would like to be managed and how you would respond. Workers are rarely self-motivated in a Theory X environment, but are more likely to be so in a Theory Y environment.

Behavior Modification

B. F. Skinner devised a theory of behavior modification that advocates positive reinforcement, namely, rewards for "good" behavior. Using Skinner's theory, one would induce people to behave differently (that is, consonant with project goals) by rewarding them when they act appropriately. "Rewards" typically are consistent with Herzberg's findings, namely, a sense of achievement and recognition.

Negative reinforcement can stop undesired behavior, but it requires positive reinforcement to promote desired behavior.

Thus, if a designer does a fine job on your project, it is appropriate to send her a memo and send a copy to her boss (or vice versa) and perhaps to her personnel folder as well. Conversely, if you have been practicing behavior modification and positive reinforcement consistently, you do not have to do anything if on another occasion she does a poor job. The absence of positive reinforcement will be message enough. Further, that absence may very well motivate the designer to ask you how her performance fell short.

Very early in my industrial career, the president of my company sent the following letter (on engraved, personal stationery) to me at my home just after the engineering project team of which I was a member shipped the first unit of an advanced system:

> Dear Milt:
>
> I would like to extend to you my hearty congratulations for your contribution to the outstanding technical success of the Satrack Program. I know that you must feel proud of being a member of the team who accomplished a marked advance in the state of the art of aspheric manufacture.
>
> Our Company is now considered to be in the forefront of this development activity, and it has been through your contributions that we have achieved this position.
>
> I know that your efforts were great and there were many long evenings and weekends which you personally sacrificed. It is indeed gratifying that we have the people with the spirit to undertake such a challenging problem and carry forth to a successful conclusion.
>
> Sincerely yours,

People's behavior can be modified.

The recipient of such a letter will become motivated to put in similar extraordinary efforts in the future and probably will continue to have family support for that effort. Other examples of positive reinforcement include awards (wall plaques, luncheons or dinners, trips, and so forth) or a story or picture in the company newspaper. These promote worker motivation because they provide recognition and validate an achievement.

Management by Objectives

Another practical technique that the project manager can use is management by objectives (MBO). Variations of this technique are management by results (MBR) and management by commitment (MBC). As the names suggest, the techniques focus management and worker attention on the outcome (objectives or results) rather than on the process by which a worker chooses to achieve the outcome. This frees the manager to concentrate on what is desired, and it leaves the worker free to concentrate on how to accomplish it in his or her own style.

To use this simple, powerful technique, the project manager negotiates with a worker on the results the worker will agree to achieve. Because the technique is a general one, the worker may be a member of the project team or the support team. The agreed results must be like project objectives, that is, verifiable, measurable, specific, and achievable. These are recorded on paper (perhaps a standard form), and are signed by both the worker and the manager. In the event that changes are required, these must be jointly negotiated, and the paper must then be revised and signed again. Note that this technique is identical to how you would contract with another organization to do subcontract work; the only difference is that the MBO agreements are not legally binding.

MBO is a powerful, useful technique.

The technique draws upon the motivational factors of responsibility and achievement. The worker may be motivated because he or she is consulted in a responsible way in defining the task to be carried out. And, as the worker is doing the work, there may be clear progress during the work that provides a periodic sense of achievement.

The only problem with the technique is that it is sometimes difficult to set the objectives or results that are to be achieved. When this is the case, the same technique used for starting a project with unclear objectives can be used. That is, agree on something specific that will be accomplished in a given period, and then try to set longer term goals when more is known about the task. Conversely, if you cannot establish a specific goal for a worker to achieve, it is clearly unfair to hold him or her accountable for something you could not describe.

STIMULATING CREATIVITY

The engineering project manager must also stimulate creativity. This is required to some extent on all projects, but it is especially applicable to high-technology projects. Even two similar engineering projects, for instance, building two identical radio telescopes, can encounter different soil conditions requiring an innovative approach in one case. In general, the less precedent for the project, the more creativity will be required.

Motivation

Creativity may be stimulated simply by managing in a way consistent with behavioral and motivational theory. Encouragement by providing recognition and appreciation is the most straightforward technique to stimulate creativity. In addition, one must provide a favorable atmosphere. In a sense, we are looking at positive reinforcement again. People are permitted to fail when asked to produce creative results and are not castigated for doing so. Rather, they are praised when they succeed.

Brainstorming

Positive reinforcement and brainstorming will stimulate creativity.

Brainstorming techniques are often used to deal with some intractable problem. The conventional method for brainstorming is to advise perhaps a half-dozen people of the problem and after one or two days convene a brainstorm meeting. At this meeting, restate the problem and reiterate the ground rules:

1. Absolutely no criticism (including smirking) is permitted.
2. The more ideas produced, the better is the session.
3. Novel, unusual (even impractical) ideas are desired.
4. Improvement or combination of prior ideas is also desirable.

Use a tape recorder to permit more leisurely subsequent consideration of the ideas thus generated.

Facilitated brainstorming is very productive.

An alternative method that works better in most situations is to have a facilitator talk to a few people individually and ask for their ideas on solving the problem. After three or four people have been interviewed, the facilitator will have a list of ideas to use to start the brainstorming session (like pump priming), which is then carried out in the normal way. This facilitated brainstorming produces more ideas and these are of better quality.

TYPICAL PROBLEMS

Project managers are not normally selected from a pool of trained, qualified people. Rather, projects arise within (or descend upon) the organization, and a person who has demonstrated technical proficiency is asked to become project leader.

Such people are often good "doers" and have technical skills and may think they want to be an engineering project manager, but they usually take the job not knowing what is involved. In general, engineers rapidly master planning techniques and then the mechanics of project monitoring. They may get along with others (as opposed to being hermitlike) but be unable to cope with the inevitable conflicts that bedevil the project manager. Or they may be poor communicators. What then happens is the organization has a poor project manager and has lost the services of a good engineer.

One cure for this problem is to be sure that candidates for project management read books such as this prior to being offered jobs as engineering project managers. After that, assuming a continuing interest in the job, the selected candidates should be offered further training, as discussed in Chapter 25.

MATERIALS STUDY PROJECT

At SUPROMAC, Ed Ross has been concerned for some time about the development of more engineering project managers and has periodically conducted in-house seminars on project management. Thus, Mel Chase, who has demonstrated competency as a mechanical engineer as well as exhibiting drive and personal ambition, attended such a seminar two years before he was asked to manage the NERESCO materials study project.

continued

However, in common with most other engineers, Mel has still not mastered many of the engineering project management skills. Thus, he remains unsure of himself in working with people. In such a situation, both he and his management must seek out further training to improve this most critical skill. Because this is often less threatening when done without the inhibition of co-workers, Mel locates a university seminar that deals with his areas of concern. He requests authorization to attend, and Ed Ross concurs and thus approves attendance at SUPROMAC's expense.

HIGHLIGHTS

There are many limits to authority; so project managers should learn how to wield influence.

Managers must confine themselves to planning and let others perform the tasks.

Human relations skills are vital to a project manager.

Familiarity with theories of motivation will help managers do their job.

Creativity can be stimulated by positive reinforcement and brainstorming.

FURTHER READING

R. D. Archibald. *Managing High-Technology Programs and Projects.* New York: Wiley-Interscience, 1976.
   Chapter 3 is an excellent overview of the many role issues confronting project managers.

V. G. Hajek. *Management of Engineering Projects.* New York: McGraw-Hill, 1977.
   Chapter 1 is only five pages long, but it has an overview of the project manager's role, stressing contractual involvement.

P. Hersey and K. H. Blanchard. *Management of Organization Behavior: Utilizing Human Resources*, 3rd ed. Englewood Cliffs, NJ: Prentice-Hall, 1977.

This is a very good and reasonably nontechnical review of motivation and organizational behavior.

F. Herzberg. "One More Time: How Do You Motivate Employees?" *Harvard Business Review,* vol. 46, no. 1 (January–February 1968), pp. 53–62.
This is a classic, brief article that summarizes job factors that are "satisfiers" and "dissatisfiers."

J. Keen. *Managing Systems Development.* New York: Wiley, 1981.
Chapter 12, pages 236–260, is devoted to people management in the data-processing environment.

P. W. Metzger. *Managing a Programming Project,* 2nd ed. Englewood Cliffs, NJ: Prentice-Hall, 1981.
Pages 129–145 provide a good but brief discussion of the manager's job on a programming project.

E. H. Porter and S. E. Maloney. "Strength Deployment Inventory—Manual of Administration and Interpretation," rev. ed. Pacific Palisades, CA: Personal Strengths Assessment Service, 1977.
This manual (designed to accompany the Strength Deployment Inventory, a personal assessment instrument) summarizes many interpersonal relationship issues.

A. M. Ruskin and W. E. Estes. *What Every Engineer Should Know About Project Management.* New York: Marcel Dekker, 1982.
Chapter 5, pages 79–118, is devoted to personnel behavior, covering many of the topics in this chapter from a different point of view. It also contains a thorough description of Porter's behavior model as it applies to engineering project management.

16

Practical Tips for Engineering Project Managers

Because an engineering project manager must be a superb communicator, I first discuss the general problem of communication and then provide several simple suggestions for improving your communication skills. Another pervasive problem for engineering project managers is resolving conflicts, and some techniques to deal with conflict are provided. I also briefly discuss how a manager can gain time to work more effectively with people. Finally, I offer some miscellaneous pointers.

COMMUNICATION

Communication must be worked at.

Effective communication is one of the more difficult human endeavors. There are so many obstacles it is amazing that any effective communication at all occurs. Words have different meanings, and people often have different perceptions or orientations. The engineering project manager's reputation (be it as a jokester or as a very serious person) will alter the way any message is received. Everyone the project manager communicates with will tend to hear the message he or she wants or expects to hear, which is not necessarily the message the project manager is attempting to deliver. Sometimes people are not listening, are distracted, or have a closed mind.

There is an aphorism about how to communicate: First, you tell people you intend to tell them; then you tell them;

and then you tell them you told them. There is much truth in this use of multiple message delivery.

There are several general steps you can take to improve your communication with other people:

Plan what is to be communicated beforehand rather than trying to decide while communicating. As it is sometimes stated, "Put brain in gear before opening mouth."

Use face-to-face meetings in which you can observe the other person's "body language." Allow enough time at an appropriate time of the day.

Decide which sequence and combination of telephone discussion, face-to-face meeting, and memo will be most effective.

Be consistent and follow through with actions appropriate to your message.

Use simple language.

In addition, you should consider using feedback, notices, and proximity.

Feedback

Communication is very much like a servomechanism in that it is not effective unless there is feedback. Communication can be improved by asking the person to whom the message has been delivered to restate it in his or her own words. This can help overcome a listener's closed mind. Another effective technique is to back up any verbal communication with a memo. This may also be done the other way around, first sending the memo and then having a meeting to discuss it. The duality of mode and the recipient's restatement, rather than simple redundancy, is most effective here.

Effective communication requires feedback.

Notices

It is impractical to meet constantly with all participants on a very large project. Even on a smaller project, it may be disruptive to have numerous meetings. It is thus desirable to issue project notices and reminders of priority actions for any given period. Putting such notices on distinctively colored paper or preprinting the project name on the top will set them apart from the other mail.

Proximity

Locating the people on the project near each other also aids communication. Because the people are close together, they can see each other more often, which makes communication easier and more frequent. And when people are in frequent contact, their point of view tends to become more uniform.

Follow-Up

It is necessary to have some system of follow-up of the communications, be they face to face or written. Some people simply keep an action log, a chronological listing of all agreements reached with other people for which follow-up action is expected.

Somewhat more effective is a follow-up system keyed to the individual from whom action is expected. A filing card with each key person's name printed on the top may be used to record notations of actions expected of that person. A variant of this is keeping a folder for each key person in which you store records of all discussions or copies of memos for which follow-up action is required or requested. In either event, hold periodic meetings with each key person and use the filing card or folder to plan the topics to be discussed.

Follow up communications.   When it is known that engineering project managers (or any manager, for that matter) have such a consistent follow-up system, people who work for or with them will realize that any statements made to them will be taken seriously. Therefore, commitments made to them will tend to receive serious and consistent attention.

CONFLICT RESOLUTION

Projects are fraught with conflicts. They inevitably arise because projects are temporary entities within more permanent organizations. One root cause is thus competition for resources. Regardless of organizational form, project managers and functional managers tend to have momentary interests that are at odds; so engineering project managers must expect and be able to "stomach" conflict. If you have a low tolerance for conflict, being an engineering project manager can be frustrating.

The Thamhain and Wilemon study (as well as one published in 1975) reviews the causes of and ways to resolve conflict. Their findings indicate that many things can be done to reduce conflicts, the simplest of which is having good plans, current and realistic schedules, and thorough communications.

The project manager must cope with conflict.

EFFICIENT TIME MANAGEMENT

Given the wide range of engineering project managers' duties (in a sense, they must be all things to all people), they can easily end up working nights and weekends unless they are very efficient in the use of time. Of course, they should not make the mistake of being efficient to the point of being ineffective. Effectiveness is achieving the desired results. Effectiveness is what counts, but project managers are more likely to be effective if they use their time efficiently.

The overriding issue in time management is "first things first." The project manager must know the most important things to do this year, this month, this week, today, and right now. Only when he or she has a clear perception of priorities can a project manager effectively manage his or her time. Thus, each day select a few (not more than three or four) really important, high impact matters for your primary attention.

The second key issue in time management is to devote large chunks of time to important single issues (for instance, one two-hour meeting with a colleague, rather than twelve fifteen-minute meetings). This can be accomplished by maintaining a time log on how you actually spend your time next week (see Figure 16-1). At the end of the week, examine the record of what actually occurred, as you recorded it at the end of each time interval during the week. Then plan how you will alter your behavior in the following week to get fewer, larger chunks of time concentrated on single topics. Do this several times over a period of several months and you will master the art of better time management.

Manage time efficiently.

Plan all meetings and always circulate an agenda in advance.

Because project management involves integrating the work of many people, numerous meetings will be held. Conducting them efficiently and effectively is essential. The following are keys to improving meetings:

> Know beforehand why the meeting is to be held and what outcome is expected. Consider whether it is possible to omit the meeting.

Time \ Day	M	T	W	T	F
8:00-8:10					
8:10-8:20					
8:20-8:30					
5:20-5:30					

Figure 16-1.   Sample time log.

Source: "Basic Management Skills for Engineers and Scientists." Lecture notes by Milton D. Rosenau, Jr. Copyright © 1982 by the University of Southern California. Reprinted by permission.

Determine the minimum number of people required.

Choose a meeting location with a room arrangement consonant with the meeting's purpose (for example, a round table arrangement for discussion among equals, a lecture hall arrangement for a presentation, and so forth).

Circulate an agenda with topic durations to all attendees, and perhaps discuss this individually with the key participants ahead of time.

Be prepared, and open the meeting on time with a restatement of the purpose and agenda.

If possible, ask each attendee (one at a time) for his or her views on each topic prior to topic completion. Just because some attendees are loud or dominant does not mean they are best qualified to speak on a topic; a quiet, shy, or retiring person will often make a valuable contribution if invited to comment.

Verbally summarize what transpires at the meeting and later distribute published minutes to all attendees.

They won't work everywhere, but you might want to try the meeting rules used on the Gossamer Albatross project (as reported in *Technology Review*, April 1981, p. 56):

All meetings are held standing in a circle.

All participants are heard in turn.

All meetings must result in a definite decision.

All decisions must be acted on immediately.

PRACTICAL TIPS

There are a few "tricks of the trade" that project managers should employ to help put the techniques discussed in this chapter to practical use.

First, keep your door open. This encourages people to talk with you, and sometimes that will identify key project issues that you were not yet aware of.

Second, close your door and do not answer the telephone. This is when you do your planning, to gain high leverage on time use. When you are also a worker on your project, which is a common situation on smaller projects, this is when you do your own technical work.

Third, walk the halls. There are always some people who will not enter your office, even if the door is open. Also, what you inevitably see when you go to the sites where work is supposed to be done is that things are not as you expected them to be.

Fourth, set a good example. Arrive early (or at least punctually) for work; take your job seriously; work hard; and be respectful of others (especially upper management and your customer), even if you disagree with some of their actions.

TYPICAL PROBLEMS

The development of your own natural style of management may be a problem when applying the practical tips in this chapter. My style of dealing with conflict or improving time management, for instance, may not be best for you.

In general, however, the biggest problem is to become a superb, versatile communicator. You must work with varied people—many (or most or even all) not of your own choosing—and that can only be done if you communicate effectively with them. Some may receive and reply to written memos, and perhaps you can write these well. Or (and this will always be the case with some workers) they may not read or understand written memos. So you must talk with and get

continued

useful responses from these other workers. Obviously, the reverse may be true also, that is, your writing skill may be relatively weaker than your speaking skill. But some workers receive written information better than spoken; so you may have to improve your writing (or reading) skills.

MATERIALS STUDY PROJECT

Mel Chase walks around to each group working on the materials study project every week, spending a full hour with each responsible task manager. (Because there are never more than three tasks running in parallel, this requires only a morning or afternoon each week.) He varies the day and time from week to week to avoid predictability. If his walk around time was predictable, some task managers might always try to avoid him and some others might try to "dress up" the appearance of their work.

HIGHLIGHTS   *Effective communication can be aided by feedback, issuing notices, and locating workers near each other.*

Conflict between different people (or their group managers) must be expected on an engineering project. Conflict can be reduced by having plans that are current and by good communications.

The engineering project manager can be more effective if he or she is a good time manager.

FURTHER READING

B. F. Baird. *The Technical Manager.* Belmont, CA: Lifetime Learning Publications, 1983.

This book is subtitled, "How to Manage People and Make Decisions," and it is a truly excellent book on this subject.

A. S. Grove. "How (and Why) to Run a Meeting." *Fortune* (July 11, 1983), p. 132.

This article provides some practical tips on how to run effective and useful meetings.

A. M. Ruskin and W. E. Estes. *What Every Engineer Should Know About Project Management.* New York: Marcel Dekker, 1982.

Chapter 5, pages 79–118, is devoted to personnel behavior, covering many of the topics in this chapter from a different point of view.

H. J. Thamhain and D. L. Wilemon. "Conflict Management in Project Life Cycles." *Sloan Management Review,* vol. 16, no. 3 (Spring 1975), pp. 31–50; and H. J. Thamhain and D. L. Wilemon. "Leadership, Conflict, and Program Management Effectiveness." *Sloan Management Review,* vol. 19, no. 1 (Fall 1977), pp. 69–89.

The first article reviews the kinds of conflict that project managers encounter and some ways to cope with it. The second article measures which influencing techniques are most effective in given situations.

N. E. Thornberry and J. R. Weintraub. "The Project Manager: What It Takes to be a Good One." *Project Management Quarterly* (March 1983), pp. 73–76.

This report on a research study (sampling 110 male high-technology project managers) identifies five core dimensions as necessary for effective performance: oral communications, influencing skills (that is, leadership), intellectual capabilities, ability to handle stress, and work skills.

Part 4

MONITORING PROGRESS ON AN ENGINEERING PROJECT

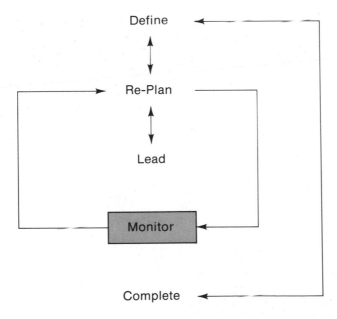

Monitoring Tools

The next managerial activity on an engineering project is monitoring progress. First this chapter discusses various monitoring techniques. Then there is a detailed consideration of the use of reports. Finally, I discuss the special case of monitoring several projects simultaneously.

CONTROLLING TO ACHIEVE OBJECTIVES

The word "control" has a pejorative connotation, implying power, domination, or authority. Thus, many engineering project managers (especially new ones) tend to avoid the necessity of installing and using controls on projects. The purpose of such project controls is to measure or monitor progress toward your objectives, evaluate what needs to be done to reach these objectives, and then take corrective actions to achieve the objectives. Thus, you must employ controls (in the measurement sense of the word) or your project will go off course and you might never know it.

Controls are needed to monitor actual progress compared to the project plan.

CONTROL TECHNIQUES

The first, and in many ways the most important, control is a well-publicized plan for all three dimensions of the Triple Constraint. A work breakdown structure, a network diagram

What you measure is that to which attention will be given.

that indicates every element of the WBS, and a cost estimate for each activity indicate how the project should be carried out. Any deviation—and there normally are several—from this three-dimensional plan indicates the need for corrective action. Without such a plan, control is impossible.

There are several restrictive control tools available, such as withholding resources or discretionary authority. When the engineering project manager uses these controls, he or she is assured that people working on the project request the use of these resources or authorities, thus providing visibility. As an example, the project manager could require any expenditure in excess of $1,000 to receive his or her specific approval. Or the project manager could require any drawing release to need his or her signature. These kinds of controls go beyond the project plan in that they make project workers seek out the project manager for approval during the performance of each project activity or task. Anyone's failure to request approval of a planned major purchase tells the project manager that the project has deviated from the plan.

Controls can tell you if the project plan is being followed.

This kind of restrictive control may well be appropriate with an inexperienced team or on a difficult project. But it is normally appropriate only for very small projects. If essentially all decisions on a large project must flow through the manager, the project will get bogged down by his or her lack of time to review a myriad of documents for approval. An effective variation on this restrictive approval control approach is to insist on independent inspection and quality control approvals or on test data as means to verify progress. For instance, you could insist that each subsystem test be approved by people working on other related subsystems.

Another project control method is to place trust entirely in the person carrying out a particular task. This method is fine if that person is able to recognize deviations from plan and realizes they must be reported promptly to the project manager. The person must also be capable of reporting the problem clearly. Because these three preconditions are rarely satisfied, this control tool should not normally be used.

Control is best exercised by examining the status of tasks.

A far better approach is for the project manager to examine the work being done under the direct control of the project team and support teams. This kind of control is based on the Theory Y assumption that people working on project tasks will be trying to do a good job (which can often become a self-fulfilling prophecy), rather than on the Theory X assumption that people will not do a good job. These examinations of activity work are accomplished by reading reports and conducting project reviews.

REPORTS

Reports fall into three broad categories: those concerned with the accomplishments along the performance axis, those concerned with schedule progress, and those concerned with cost. Reports may be written as summaries to provide an overview or be detailed about a particular task activity or some other element of project work. Reports may be strictly for the use of the organization performing the project or be intended for people outside, such as the customer or contractors. If the project organization is an industrial company, such as SUPROMAC, there is a wide spectrum of people to whom reports might be addressed, as shown in Figure 17-1.

Problems

The project manager cannot depend entirely on reports. In the first place, they may be inaccurate, a common failing of cost reports, which are prone to arithmetical errors and data entry errors if a computer-based system is being employed. On the other two axes, the people who write reports are prone

Do not confuse activity with progress.

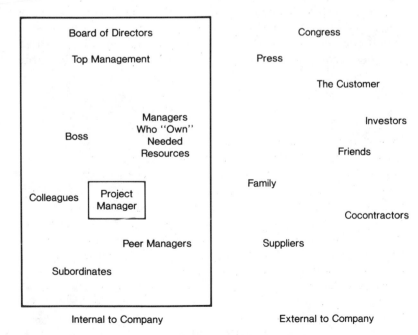

Figure 17-1. Report recipients.

to unwarranted optimism. Project workers generally assume a well-advanced task is nearly complete; in fact, no task is complete until it is truly finished. Thus, most task reports will indicate a task is 80 or 90 percent complete, implying it will require additionally only a small fraction of the time already spent. Unless the task is as simple as drilling a hundred holes in a plate, it is normally impossible to measure what percentage of the task is truly complete. (And, as I said before, even in this case, there is no assurance the last hole won't destroy the plate, forcing the task to be completely redone.) In general, completing any given task requires more time than the person working on it forecasts. For instance, a report that X engineering drawings have been completed for an engineering project originally planned to require Y engineering drawings tells you nothing about the percentage of completion. You know you are truly complete only when Z engineering drawings (which may be either more or less than the Y drawings planned) actually are completed, checked, and released.

Even though the task manager's report expresses a high degree of confidence that the task will be completed when scheduled, such a promise is not deliverable. Reports are one-way communication and lack the give and take possible in a meeting; so it is hard to judge the status of a task in progress solely from written data.

Consider the schedule situation illustrated in Figure 17-2. There is a long period of time in the middle of the project when there are no scheduled task completions. Thus, there are no certain checkpoints available, leaving considerable

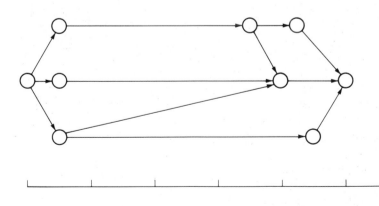

Figure 17-2. Time-based network diagram.

uncertainty as to the actual status. This illustrates another reason to break a project into many small tasks because there will be less uncertainty as to schedule status.

If the work breakdown structure divides the engineering project into many small tasks, the manager can look at each of them individually and decide whether they are complete. Only those tasks for which all work has been done are considered complete. No other task, regardless of the amount of effort applied so far, is complete. A task that is 80, 90, or 99 percent complete is not complete. This two-state (bivalued) approach to examining tasks simplifies the engineering project manager's job tremendously. He or she can accept as complete only those tasks for which reports (oral or written) guarantee the activity is complete. In fact, the task authorization document (Figure 6-3) could have additional blocks for signature by the task manager and project manager to indicate their agreement that the task is satisfactorily complete (see Figure 17-3). With many small tasks rather than only a few larger ones, there is much less uncertainty about overall project status.

Dividing a project into many small tasks aids precise determination of its status.

Detail Level

Reports should always be as brief as possible. Because many people concerned with the project will wish to receive reports, there is a tendency to try to circulate one report to many recipients. This is a mistake. Many who wish to be kept abreast of progress are not interested in small details. This is especially true of busy management, both project organization and customer, for whom the engineering project manager should prepare brief, summary level reports. Such reports might also be circulated to people concerned with some specific aspects of the project but who do not require all the details in any particular report.

Pictures, demonstrations, and models should also be encouraged, especially for a high-technology project or one where much of the work is geographically remote. It is often hard for people not intimately concerned with the project to visualize the status, expected outcome, or even the concept. For them, tangible descriptions, pictures, and such are by far the most appropriate means for providing reports. If your organization has closed-circuit television, this may be an excellent use of it.

It is always important to avoid the pitfalls of excessive reporting. Clearly, this is a gray area requiring judgment. The "convenience" of copying machines can easily lead to too many copies of overly detailed reports being circulated to too many people.

Cost Reports

Cost planning and reporting are important because they provide a guide for management action. They permit com-

TASK AUTHORIZATION	PAGE OF	
TITLE		
PROJECT NO.	TASK NO.	DATE ISSUED
STATEMENT OF WORK:		COMPLETION REPORT:
APPLICABLE DOCUMENTS:		
SCHEDULE START DATE:		COMPLETION DATE:
COST: PLANNED =		ACTUAL =
ORIGINATED BY: DATE:		ACCEPTED BY: DATE:
APPROVED BY: DATE:		APPROVED BY: DATE:
APPROVED BY: DATE:		APPROVED BY: DATE:
COMPLETED BY TASK LEADER:		ACCEPTED BY PROJECT MANAGER:

Figure 17-3. Task authorization form modified to serve also as a task completion report and acknowledgment form.

parison of actual accomplishments in terms of cost incurred and planned accomplishments. Thus, they aid in determining whether there will be a cost overrun or underrun at completion. In some instances, they may aid in determining whether a specific task is on schedule.

Unfortunately, in many organizations, there is a tendency to look at actual cost versus planned cost as virtually the only measure of project progress. This is useless because there may be offsetting overruns and underruns, which I discuss in the next chapter.

MULTIPLE PROJECTS

As an engineering project manager, you may eventually be responsible for more than one project. You may have many project managers reporting to you or there may be so many task managers reporting to you on a single project that you cannot personally attend all the task reviews and critically examine all the necessary detail. In this case, you must receive some kind of summary information that indicates the status of the several projects (or the many tasks) for which you have responsibility.

When you cannot personally get into details, you are completely at the mercy of those who summarize information for you. Robert A. Howell indicates one way you might receive summary briefing information. In essence, his proposal is that, for every project (or task) for which you have responsibility, there should be a red, yellow, or green status indicator for each of the technical, schedule, and cost dimensions of the job. He also suggests these for funding status and would display in this summary form whether project review meetings have been held and at what point the program plan obtained approval. Such a system is certainly graphic, but it is no better than the judgment of the person who prepares the information.

The value of summary status indicators is questionable.

Probably a better approach is to visit the various project reviews personally. Imagine you have three engineering projects reporting to you, each of which is being managed by another project manager, who reviews it monthly. You might sit in on each of these in a rotating fashion; so you attend the review of each project once every three months.

Personal reviews are best.

TYPICAL PROBLEMS

Problems arise because of too much or too little monitoring. The former may demotivate personnel, consume too much time, or cost too much. Inadequate monitoring, based perhaps on the naive optimism that the project will be performed in accordance with plan, can lead to disaster. The following (edited) quotation comes from a memo written months after the start of an engineering project:

A. The accessories were far more complex than anticipated at the time the original plan was prepared. There was not, at that time, any defined level of performance for these accessories; the feasibility of the accessories had not been studied. The estimates for the optical systems and mechanical components of the accessories were inadequate until several months of feasibility study had been performed and the products fully defined. Even after they were defined, all of the accessories have had to go through repeated design iterations. In October, all of the accessory designs were rejected. The Head Adapter and the Lamp Housing required mechanical design rework to reduce their size. In addition, the Lamp Housing required optical redesign to accommodate additional optical features necessary for a smaller, more compact package. The Head Adapter was finally approved in mid-December; however, the base was not approved until mid-January. The Lamp Housing received verbal approval only last week. The Accessory, as designed in this project, was found to be completely unacceptable, requiring total redesign, forcing its removal from this project (all future efforts on the Accessory will be covered by a separate project and are excluded from this summary).

The original plan of last February (13 months ago) contained approximately 600 hours of optical and mechanical design for the Accessories. In our revised plan, including labor to date, the optical and mechanical design is approximately 2400 hours; the difference is 1800 hours plus drafting time.

B. The original plan called for 670 hours for industrial design. Because the designs have been subjected to repeated revisions, the revised plan, includ-

continued

ing labor hours spent to date, puts industrial design at approximately 1700 hours: an increase of 1030 hours.

C. The layout and design of the Indicator Dial required considerably more time than that provided in the original plan. The Dial design was rejected several times, which added approximately 800 hours.

D. The custom microcircuit for the digital meter version added coordination time with the vendors and with purchasing of approximately 300 hours.

E. It was assumed in the original plan that the proprietary microcircuit would be available as an input to this project. Delays in receiving properly functioning devices made the timing critical. Therefore, coordination with the vendor and evaluation of sample devices of about 400 hours was performed on this project.

F. Due to the complexity of the mechanical design, we added preparation of a tooling plan, preliminary tool design, and coordination to the project. This has added approximately 450 hours. (We will also do the tool design, but that work is covered under a separate project.)

G. The revised plan contains 90 hours for planning and supervision; that time is now charged to the project rather than handled as indirect overhead.

H. The original plan was based on burdened labor rates that were in effect when the plan was prepared. The change in burdened labor rates since then has added approximately 20 percent to the project cost.

I. Personnel being shifted temporarily to other projects has caused approximately 200–300 hours to be added.

J. Approximately 340 hours were added to support engineering time to better match experience on other projects.

K. The original plan allowed for about 800 hours for project administration. The actual and planned hours now represent an increase of approximately 800 hours due to slightly higher monthly hours plus the longer duration of the project.

L. Compared to the original plan, a higher level of technical documentation hours have been expended. Additional drafting time associated with the Accessory mechanical designs and redesigns (due to size

continued

rejections) is approximately 600 hours. Other factors were the low estimates for standard parts drawings and reworking circuit layouts due to mechanical and size limitations. These other factors represent approximately 1200 hours.

This project has gotten into severe cost (approximately quadrupled) and schedule difficulties, and it basically got that way by slipping one day at a time. Continuous monitoring could have detected deviations from plan in time to redirect—or terminate—the project.

In addition, there is a monitoring problem because cost (and other) report time lags delay news of project difficulties. Further, there are inaccurate reports, even from conscientious people. In other cases, unclear reports will mask the deviation that has actually occurred.

Also, even if the reports are prompt, accurate, and clear, the deviation may be noticed when you are busy with other urgent activities. Whenever a deviation is noticed, it takes time to react. Finally, you may be distracted by a human tendency to search for a guilty party on whom to blame everything.

MATERIALS STUDY PROJECT

Assume that Mel Chase has to decide how to control the materials selection and procurement tasks (tasks A and B) of the materials study project. Because much of the work will be routine, he opts to trust the task manager, Cheryl Williams. A discussion with Cheryl has made Mel confident that the three critical conditions are satisfied: She can recognize if there is a problem. She knows a deviation must be reported promptly. She both writes and speaks clearly. As further assurance and control, Mel asks Cheryl also to provide a brief written note on work status each month.

At the conclusion of three weeks, Cheryl reports that she has concluded task A on schedule, using the
continued

task authorization (Figure 17-4). Assuming Mel Chase
agrees that the task is complete, he would also sign the
form and make copies for the project file and any other
people who have to be notified.

TASK AUTHORIZATION	PAGE *1* OF *1*
TITLE *Select Materials*	

PROJECT NO. 83627	TASK NO. 83627.01	DATE ISSUED 3 JAN 84

STATEMENT OF WORK:	COMPLETION REPORT:
Select specific test materials that conform to our proposal.	*Have chosen X and Y as meeting requirements of NERESCO, and satisfying our need for price and availability.*
APPLICABLE DOCUMENTS: *Proposal (attached)*	

SCHEDULE

START DATE: 3 JAN 84	COMPLETION DATE: 20 JAN 84
COST: PLANNED = *$5,250.—*	ACTUAL = $ *$6,850.—*
ORIGINATED BY: DATE:	ACCEPTED BY: DATE:
APPROVED BY: DATE:	APPROVED BY: DATE:
APPROVED BY: DATE:	APPROVED BY: DATE:
COMPLETED BY TASK LEADER:	ACCEPTED BY PROJECT MANAGER:

Figure 17-4.   Task completion report for task A of materials study
project.

HIGHLIGHTS   *Comparison with the project plan provides the basis for monitoring and controlling an engineering project.*

Project managers may exercise control by requiring their approval or trusting task managers, but the best approach is to examine the status of tasks.

Reports, which may concern any axis of the Triple Constraint and be detailed or general, provide a means to examine the status of tasks.

Managers in charge of several projects can best exercise control by periodic personal reviews.

FURTHER READING

R. D. Archibald. *Managing High-Technology Programs and Projects.* New York: Wiley-Interscience, 1976.
  Chapters 8 and 9 provide several illustrations of control techniques.

R. H. Clough and G. A. Sears. *Construction Project Management,* 2nd ed. New York: Wiley-Interscience, 1979.
  Chapters 9 and 10 stress time and cost control, also illustrating several reporting techniques.

R. A. Howell. "Multiproject Control." Managing Projects and Programs Series. Cambridge, MA: Reprint from Harvard Business Review (no. 21300), 1971.
  This article clearly illustrates a method for controlling several projects simultaneously. As noted in this chapter, there are some problems with Howell's approach, but it is still the best concept if there are several projects.

R. Pilcher. *Appraisal and Control of Project Costs.* London: McGraw-Hill, 1973.
  Chapters 11 and 12 provide a detailed discussion of cost controls.

J. D. Wiest and F. K. Levy. *A Management Guide to PERT/CPM,* 2nd ed. Englewood Cliffs, NJ: Prentice-Hall, 1977.
  Chapter 6 is a brief discussion of PERT/COST as a control tool.

18

Reviews of Engineering Projects

There are two kinds of engineering project reviews: periodic (typically monthly) and topical. Reviews are the most important control tool available to the project manager; they assure that he or she and the workers will actually be meeting to discuss progress. This chapter deals with both types of reviews and the general necessity to conduct reviews.

THE NECESSITY FOR REVIEWS

Having reviews is very much like having a navigator on an airplane. Reviews and a navigator are unnecessary if everything is proceeding according to plan. The purpose of both is to uncover deviations and correct them. Experienced project managers know the project will not proceed as planned, but they do not know how it will deviate. Only the naive engineering project manager believes the plan is sufficient and no further navigation is necessary to arrive at the project's Triple Constraint point destination.

The project manager's boss and other senior management will frequently want to know about the project status. Although this may not be true for relatively insignificant projects within an organization, it is a very common situation for commercial development projects where company money is at stake and for larger engineering projects being performed on contract for others. The customer (in external projects), too, may wish to have periodic or topical reviews.

Reviews are your off-course alarm.

These requirements and their nature and thoroughness should have been included in the negotiated contract for the work. The people working on the project will also wish to have reviews of the overall project from time to time. This is their means of learning whether their effort has to be adjusted from plan to conform with some new reality or everything is still proceeding according to the original plan (which never occurs). Reviews with the project personnel in attendance are a means for communication and can enhance their motivation.

THE CONDUCT OF REVIEWS

Whether reviews are periodic or topical, they should be planned. Certain questions are almost always appropriate to raise.

Planning

Think of a review as a small project.

Reviews may be thought of as a very small project. The goal is acquisition of all relevant information. There is a schedule and cost. The schedule may be a simple statement that the review will consume two hours on a particular afternoon. The cost depends on the number of people who participate and the preparation time.

There should be a plan for reviews. Everyone involved should understand the Triple Constraint for the review and be prepared to carry out their assignments. This means the project manager must make specific assignments to various individuals, who must know how much time and what level of detail are appropriate for their participation. Reviews may also be thought of as a particular kind of meeting. As such, all the care of preparation and follow-up discussed in Chapter 16 is relevant.

Questions

Ask "why" questions.

The smart engineering project manager learns to ask questions at project reviews. You are not asking questions to embarrass or pillory anyone but rather to find out how the project is deviating from the plan so you can take corrective action. Thus, ask questions nonthreateningly. If you (or others, such as configuration management or quality assurance) don't ask questions, some people won't volunteer critical information. Such nondirective questions often start with "why." A very good question is, "Why are you doing that?" You can follow this with successive "why" questions. Some other helpful questions to ask are the following:

What persistent problems do you have, and what is being done to correct them?

Which problems do you anticipate arising in the future?

Do you need any resources (people or things) you do not yet have?

Do you need any information you do not have now?

Are there any personnel problems now or that you anticipate?

Do you know of any things that will give you schedule difficulties in completing your task? If so, what are they?

Is there any possibility your task will be completed early?

Will your task be completed within the allowed budget, or do you anticipate some overrun?

Is there any possibility your task will be completed with an underrun?

Is there any possibility that completion of your task will lead to any technological breakthroughs for which patents might be appropriate?

Has any work done on your task led to any competitive edge we might use to gain other business elsewhere?

The thing to remember about project reviews and these questions is that you are almost assuredly going to hear some bad news. Most of us do not cope with bad news in a very positive way; so the project review can easily become a recrimination and blaming session. This will not be productive. It will destroy the review and much additional effort on the project. Be businesslike and factual in conducting the reviews, and keep asking questions to gather information. If it is appropriate to assess blame, do that in a different meeting, preferably privately with the person who must be blamed.

Plan to conduct project review meetings and expect problems to surface.

PERIODIC REVIEWS

In general, every project should be reviewed once a month. It may be appropriate to review some projects once a year or once very three months; others may require weekly or even daily reviews. Periodic reviews can catch deviations from plan before they become major disasters. In the case of the overrun engineering project discussed in the previous chapter, periodic project reviews could have caught the deviations at the end of one or two months, when something might have been done about them, rather than at the end of thirteen months, when the accumulated deviation from the plan was so severe.

Task Review

A task is in one of two conditions: complete or not complete. For tasks whose performance axis dimension has been completed, examine the actual versus planned cost and schedule, as illustrated in Figure 18-1. Unless there is something unique about the cost deviations on any completed activity, the accumulated actual cost versus plan can be used to project the cost at the end of the project. In Figure 18-1, actual cost for the five complete tasks ($42,000) is less than planned cost ($44,100), and the ratio of these indicates that

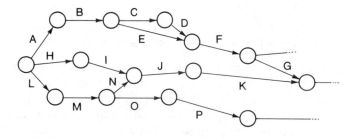

1 - Identify technically complete activities.
2 - Examine actual versus planned cost and schedule:

Completed Activity	Cost			Schedule		
	Act	Plan	Var	Act	Plan	Var
A	3,000	3,200	200	7	6	⟨1⟩
B	4,100	2,900	⟨1,200⟩	9	5	⟨4⟩
C	6,000	8,400	2,400	6	11	5
D	9,700	12,600	2,900	12	17	5
E	19,200	17,000	⟨2,200⟩	24	23	⟨1⟩

Figure 18-1. Measuring progress.

the final actual costs will be approximately 97.5 percent of the plan.

Actual versus plan ratio may not be meaningful in the case of schedule variations because many of the completed tasks will not be on the critical path. An activity not on the critical path will often be completed later than plan simply because it was not necessary to complete it within the planned time. Thus, the project manager can make predictions about the schedule only by looking at completed tasks on the critical path and then redetermining it.

Watch the schedule and recalculate the critical path.

It is appropriate to ask which incomplete tasks are in progress and which have not yet been started. For those under way, find out whether there have already been any difficulties that would preclude their being completed on time within the cost plan.

In the case of the project illustrated in Figure 18-1, the next concern would be the status of tasks H, L, and F. Do these tasks not yet completed indicate that the project is hopelessly behind schedule? After exploring that, we want to know about the critical path for this particular project. In this case, the concern is whether the completed activities have already caused the project to slip or whether the schedule variations have no significance with regard to the overall project.

In the case of cost review, it is necessary to examine the details of each task individually, as shown in Figure 18-2. The

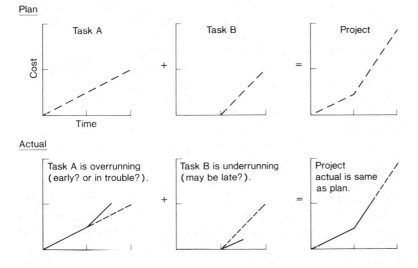

Figure 18-2. Details matter.

project manager who looks only at the overall project may be deceived. Project cost can appear to be in harmony with the plan, but that may conceal compensating task overruns and underruns or other difficulties that must be explored. Naive managers assume that actual cost in excess of plan is bad (task A, Figure 18-2), but that may only signify work being completed early. Similarly, actual cost underrunning plan is naively assumed to be good news (task B, Figure 18-2), but this may actually be symptomatic of a situation in which actual progress is inadequate. To determine real situation, the engineering project manager must ask questions, not just look at project cost estimates.

Commitments

Project cost reports are always plagued by certain problems. First, they are not issued instantly at the end of the month. It takes time to gather the data, process it, print it out, and distribute it, whether the system is manual or electronic. Second, these reports are never the highest priority in any corporation's accounting department, coming after payroll and usually after customer billing. Thus, project cost reports typically follow completion of the monthly period by two or three weeks (at least a week and a half, and, in some cases, five or six weeks).

Commitments are future costs.

In addition, cost reports are nothing but a record of the apparent charges to a project, which may be in error. Even if the cost charges to the project are correct, they do not necessarily reflect the corresponding amount of performance accomplishment. Even if the actual costs for each task agree completely with the plan for each task, there is no assurance that the work accomplishment has a corresponding degree of completion. Within an individual task, where there is no detailed further subtask breakdown, the project manager has only judgment to guide whether progress and cost are consistent. But if a project is divided into many tasks, the ambiguity is reduced because the project manager can look at completed tasks and compare their actual cost with plan. In looking at these costs for tasks, the manager must also look at commitments. Commitments are obligations for which the project will be billed but that are not reflected on the actual cost reports. Figure 18-3 illustrates two common problems.

A task apparently completed with an underrun may in fact have outstanding bills charged against it. To preclude surprises, look at the commitments charged against the task.

You may do so with either a manual or computer-based commitment report. If these reports do not exist, the project manager must maintain these manually to avoid unpleasant surprises.

To summarize, review the Triple Constraint:

- Technical dimension
  Task is either complete or not.

- Schedule dimension
  Compare actual and planned time for completed tasks. Slippage is significant only if the schedule is delayed.

- Cost dimension
  Compare actual and planned cost for completed tasks. Include commitments.

Cost to Complete

Periodically, the engineering project manager should request estimates of the cost (and time) to complete all incomplete tasks. You might do so quarterly or semiannually and also whenever the plan is revised substantially.

Do not make these estimates by simple subtraction of cost to date from the planned cost, as in method 1 in Figure 18-4. Have each task manager totally reestimate the task in light of everything now known, as shown by method 2 in the figure. Unfortunately, this often indicates that the project will overrun its budget, but it is important to learn of that possibility

Month	Event	Commitment Report	Cost Report
1	$10K purchase order issued	$10,000	-0-
2	$1K trip authorized	11,000	-0-
3	Traveler paid $900	10,000	$900
4	Vendor delivers	10,000	900
5	Vendor sends bill	10,000	900
6		10,000	900
7	Vendor's bill paid	-0-	10,900

Figure 18-3. Commitments are future costs.

early enough to do something about it. As the project progresses, the participants learn about what is involved. Also, replanning may occur, and the necessity for new tasks may become apparent.

Follow-up Actions

During any review, a variety of actions will be identified to cope with the various problems uncovered. The project manager should always record these actions, the person responsible, and the expected completion date. This might be done on a register (Figure 18-5). All concerned people should receive copies, and the status of these action assignments should be reviewed no later than the next project review.

TOPICAL REVIEWS

There are many kinds of topical reviews for engineering projects. The one used depends on the undertaking and customer requirements. Many hardware projects have the following types of reviews:

1. Conceptual design
2. Preliminary design
3. Critical design
4. Manufacturing
5. Preshipment
6. Management
7. Customer

Method 1		Method 2	
Plan (or Allowed) Cost	= $10,000	Task Q	= $1,632
Cost to Date	= 6,000	Task R	= 2,149
Cost to Complete	= $4,000	Task S	= 1,910
		Cost to Complete	= $5,691

Figure 18-4. Estimated cost to completion.

A software project would typically have different types of reviews, as would a construction project.

All these reviews are often mandated in the contract and may be a precondition to carry out further work on the project. In such a case, the reviews themselves would clearly be designated as task activities on the network diagram and be considered major project milestones, as discussed in Chapter 7.

Management and customer reviews are often onerous, but they may stimulate participation and involvement on the part of all people working on the project. For this to happen, the project manager must solicit ideas from all project per-

Assigned Actions for Completion Project: _____		
Action	Responsibility	Due

Figure 18-5. Typical action follow-up form.

sonnel as to what should be discussed during the review. Rough ideas for the review should be delineated in a smaller group of key staff. At this point, the manager should delegate portions of the review to other people. (Be sure there are not so many people making presentations at the formal review that it becomes a circus.) Next conduct a trial run with a fairly large group of project participants invited to criticize. Following this, prepare materials for the formal presentation, before which it is desirable to conduct a second dry run with a peer management group that represents the same range of skill backgrounds as the audience that will attend the formal review. The people attending this dry run review will provide additional insights as to how materials can be better presented or changed. If possible, videotape the presentation for later review.

Major review presentations can be an opportunity.

Now you are ready to conduct the formal review, after which it is desirable to conduct the entire review again for the benefit of everyone working on the project. They are just as interested as you in the kinds of questions that management and the customer asked and will find it just as interesting and motivating to spend some time hearing what happens and getting an overview of the particular issues covered in the review.

TYPICAL PROBLEMS

Reviews are plagued by two common problems. First, there is always a concern as to whether the information being presented or discussed is accurate. Cost reports, as previously noted, are especially prone to error. Beyond this, there is often speculation about the exact status of some task, test, component, or report. Clearly, good planning for reviews can reduce if not eliminate this problem.

The second problem is the poorly conducted review. Aimless discussion or recriminations are common. Running a review like any other well-planned meeting greatly reduces the possibility of getting off the track.

MATERIALS STUDY PROJECT

At the fourth monthly project review, the summary information in Figure 18-6 is presented to Mel Chase. All tasks that are supposed to be completed have actually been completed, as has task F, and progress is being made on the two other tasks. In addition, actual costs to date for completed tasks are below plan. But, although there has been no schedule slippage of tasks on the critical path, a careful examination of schedule shows that the original critical path has been replaced by a new critical path (the rest of task H, plus tasks J and K), due to the very late completion of task G.

Consider another possibility. Imagine that SUPRO-MAC management requires a preshipment review before the final report (task K) can be sent to NERESCO. Because this is not shown explicitly as a separate task, it must be integral in the task (that is, comparable to typing, editing, binding, and so forth). Thus, when the project enters the last month, Mel Chase might prudently begin to schedule the appropriate managers' time to participate in the review.

Reviews uncover the inevitable deviation from plan and allow a consensus as to the needed corrective action.   **HIGHLIGHTS**

Reviews, like projects, must be planned.

Ask nondirective questions and expect problems to surface at project review meetings.

Periodic reviews should be conducted as appropriate for the project, but once a month is a good rule of thumb.

The kind of topical review used depends on the project and the customer's requirements.

Questionable accuracy and poor procedures are common problems with reviews.

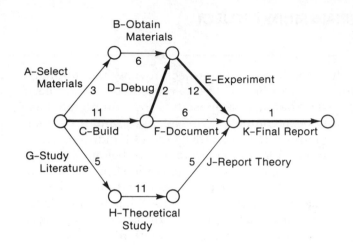

TASK	COST ($)			SCHEDULE (WEEKS)		
	Plan	Actual	Variance	Plan	Actual	Variance
A *	5,250	6,850	<1,600>	3	3	——
B *	9,125	8,725	400	6	8	<2>
▷ C *	87,270	89,890	<2,620>	11	11	——
▷ D *	10,900	10,900	——	2	2	——
▷ E	177,200	P		12	P	
F *	92,400	81,200	11,200	6	5	1
G *	18,220	19,300	<1,080>	5	10	<5>
H	42,020	P		11	P	
J	17,500	N		5	N	
▷ K	14,950	N		1	N	
TOTALS FOR COMPLETED TASKS	223,165	216,865	6,300			

Legend

▷ Critical Path
* Task Complete
P Task in Progress
N Task not Started

Figure 18-6. Summary of material study project after four months. Cost projection is for 3 percent underrun, but critical path has now changed and project will be one week late unless something changes.

FURTHER READING

There does not appear to be any extensive, systematic discussion of reviews per se as a control or monitoring tool, although the following books include brief mention and limited further treatment:

B. N. Abramson and R. D. Kennedy. *Managing Small Projects.* Redondo Beach, CA: TRW Systems Group, 1975.

R. D. Archibald. *Managing High-Technology Programs and Projects* (Chapter 9). New York: Wiley-Interscience, 1976.

V. G. Hajek. *Management of Engineering Projects* (Chapter 14). New York: McGraw-Hill, 1977.

M. Zeldman. *Keeping Technical Projects on Target.* New York: AMACOM, 1978.

19

Project Cost Reports

Actual project costs must be measured to control the cost dimension and may reveal schedule or performance dimension problems. I discuss these topics and include typical examples of cost reports, which are normally generated as computer printout, in this chapter.

COMPUTER COST REPORTS

Large organizations commonly have computer-generated reports that summarize project cost. In some cases, they may also cover schedule deviations. There is a variety of report systems available for purchase or lease. Figure 19-1 shows how such reports can be useful for project monitoring. Actual cost (and schedule) data are collected from labor time cards, purchase orders, and other direct charges to the engineering project. These are compared with plans; noted variances are analyzed; and required corrective actions may then lead to replanning. Comparing reality to the plan may suggest that certain trends will lead to future variances, which again is a cause for replanning. To be useful, this comparison must be done for each cost center (for example, department) and work breakdown task.

The project manager must steer a careful line between having too many forms and too much information and having too little information to monitor the project, although it is probably better to have more information than less. Neither the engineering project manager nor top management should be trapped into believing that actual cost data (which can be reported with great precision) are the only measure of project health (which requires difficult three-dimensional measurements).

Variance from plan is a danger signal.

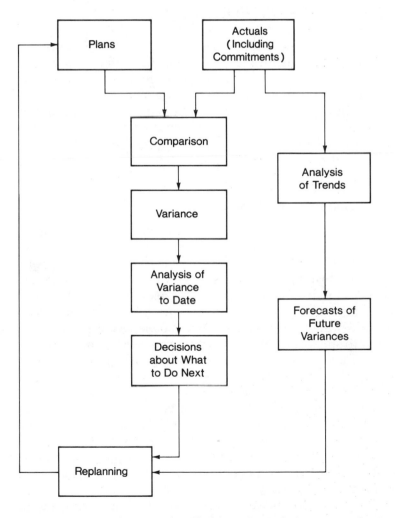

Figure 19-1. Replanning is called for whenever future trends or past actuals indicate significant deviation from plan.

Most organizations with computer support for engineering project management issue a weekly labor distribution report. This report lists the names of people charging time to each project during the previous week. This key report for engineering project managers provides an early warning signal. Examination of the report may reveal that people are charging your project who should not be charging it and that people whom you expected to be working on it are not charging your project.

COST MONITORING PROBLEMS

This section reviews a few of the cost reports for the materials study project, illustrating some issues in engineering project control. These reports are typically prepared with a computer, working with the planning data base previously discussed. But it is not absolutely necessary to use a computer, although the amount of data to be handled can otherwise be a substantial burden, even to a well-qualified project cost accountant.

Variances Due to Timing

Figure 19-2 is the project cost report for task B within the chemical group at the end of the first month. Typically, this report would be available about the middle of the second month (February, in this case). The planning numbers are consistent with Figure 9-11. In this case, labor and the overhead thereon are in accordance with plan. But there is a favorable variance in the nonlabor expense. That is, there was a plan to spend $4,400, but nothing has been spent.

A variance between actual cost and plan may be due to payments being made later than plan rather than to work variances.

Figure 19-3 shows the corresponding task and period commitment report for the chemical group. Commitments in the amount of $4,400 have been incurred. Thus, the cost variance merely indicates that bills have not yet been paid rather than being a variance due to activities not yet undertaken. To put it another way, it is anticipated that the $4,400 expense will occur later. This demonstrates that it is impossible to make intelligent use of project cost reports without also examining commitment reports.

There may also be variances caused by purchasing a large amount of some material, some of which is needed for an early task and the rest for a later task (for instance, to obtain a volume discount). In this situation, the receipt (and cost) of material do not conform to task usage.

Figures 19-4 and 19-5 are the same reports for the end of the second month. Once again, the labor hours and costs as well as the overhead are in accordance with the plan. In this

PROJECT COST REPORT

PROJECT_____ MATERIALS STUDY _____ TASK___ B-OBTAIN MATERIALS __ DEPARTMENT___CHEMICAL____

CATEGORY		MONTH 1 - JAN			TOTALS TO DATE		
		PLAN	ACTUAL	VARIANCE	PLAN	ACTUAL	VARIANCE
HOURS	SR. PROF.	40	40	0	40	40	0
	JR. PROF.						
	SR. TECH.						
	JR. TECH.						
DOLLARS	LABOR	1,000	1,000	0	1,000	1,000	0
	OVERHEAD	1,000	1,000	0	1,000	1,000	0
	NONLABOR	4,400	0	4,400	4,400	0	4,400
	PRIME COSTS	6,400	2,000	4,400	6,400	2,000	4,400
	G & A	960	300	660	960	300	660
	TOTAL COSTS	7,360	2,300	5,060	7,360	2,300	5,060

Figure 19-2.   Cost report for first month for task B work by the chemical group.

PROJECT COMMITMENT REPORT

MONTH____1____

PROJECT Materials Study TASK Obtain Materials DEPARTMENT Chemical

Commitment Date	Item	Amount	Estimated Payment Date
21 Jan 84	P.O.—Material Supply Co.	$4,000.	30 Apr 84
21 Jan 84	Travel Auth.—C. Williams	400.	15 Feb 84
		$4,400.	

Figure 19-3.   Commitment report for first month for task B work by the chemical group.

case, there is an unfavorable variance during the second month with regard to nonlabor costs because the travel voucher is paid now but the plan had the expense in the first month. For the totals to date, that is, through the end of the second month, the nonlabor variance is favorable. This favor-

PROJECT COST REPORT

PROJECT_____MATERIALS STUDY_____ TASK__B-OBTAIN MATERIALS__ DEPARTMENT__CHEMICAL__

CATEGORY		MONTH 2 - FEB			TOTALS TO DATE		
		PLAN	ACTUAL	VARIANCE	PLAN	ACTUAL	VARIANCE
HOURS	SR. PROF.	4	4	0	44	44	0
	JR. PROF.						
	SR. TECH.						
	JR. TECH.						
DOLLARS	LABOR	100	100	0	1,100	1,100	0
	OVERHEAD	100	100	0	1,100	1,100	0
	NONLABOR	0	350	(350)	4,400	350	4,050
	PRIME COSTS	200	550	(350)	6,600	2,550	4,050
	G & A	30	83	(53)	990	383	607
	TOTAL COSTS	230	633	(403)	7,590	2,933	4,657

Figure 19-4. Cost report for second month for task B work by the chemical group.

PROJECT COMMITMENT REPORT

MONTH_____2_____

PROJECT Materials Study TASK Obtain Materials DEPARTMENT Chemical

Commitment Date	Item	Amount	Estimated Payment Date
21 Jan 84	P.O.—Material Supply Co.	$4,000.	30 Apr 84
21 Jan 84	Travel Auth.—C. Williams	All	Paid
		$4,000.	

Figure 19-5. Commitment report for second month for task B work by the chemical group.

able variance is composed of the unpaid $4,000 purchase order, which is variance only because of payment timing, and a $50 favorable variance because the travel voucher payment was $50 less than plan.

Variances Due to Actual Work Not As Per Plan

Figures 19-6 and 19-7 are the project cost reports at the end of the third and fourth months for the same task. In this case, labor hours exceed plan in the third month, with attendant unfavorable cost variances, which happen to offset exactly the previous favorable variance on nonlabor due to the travel variance. In the fourth month, the purchase order is paid, and the net variance for the task becomes zero.

Figures 19-2 through 19-7 illustrate that variances occur because of the payment timing and that actual performance differs from plan. They also indicate the necessity of examining the details in project cost reports and commitment reports to understand the reported variances and their significance.

Actual work may differ from plan.

PROJECT COST REPORT

PROJECT_____MATERIALS STUDY_____ TASK___B-OBTAIN MATERIALS__ DEPARTMENT____CHEMICAL____

CATEGORY		MONTH 3 - MAR			TOTALS TO DATE		
		PLAN	ACTUAL	VARIANCE	PLAN	ACTUAL	VARIANCE
HOURS	SR. PROF.	2	3	⟨1⟩	46	47	⟨1⟩
	JR. PROF.						
	SR. TECH.						
	JR. TECH.						
DOLLARS	LABOR	100	125	⟨25⟩	1,150	1,175	⟨25⟩
	OVERHEAD	100	125	⟨25⟩	1,150	1,175	⟨25⟩
	NONLABOR	0	0	0	4,400	350	4,050
	PRIME COSTS	200	250	⟨50⟩	6,700	2,700	4,000
	G & A	30	38	⟨8⟩	1,005	405	600
	TOTAL COSTS	230	288	⟨58⟩	7,705	3,105	4,600

Figure 19-6.   Cost report for third month for task B work by the chemical group.

Each computer-based system differs in detail; so the project manager should understand exactly how the reports are prepared (that is, to what errors the reports are prone) as well as the specific meaning of each column of data.

Variances Due to Overhead Rate Changes

Figure 19-8 is the project report for task E in the technical support section, as reported at the end of the fourth month. In this report, there are favorable variances in labor but an unfavorable variance in overhead. How can this be? If labor is favorable, why should overhead be unfavorable?

Factors outside the manager's control may cause costs to vary from plan.

Figure 19-9 shows the cause of this. A planned overhead rate (namely, 100 percent) was based on a planned work load for the technical support section (or perhaps the entire research and engineering division). But the actual overhead rate at the end of the fourth month is higher (namely, 130 percent) because the work load base for the entire section has been reduced from plan. The overhead expenses have been somewhat reduced, but not in the same proportion as the

PROJECT COST REPORT

PROJECT_____MATERIAL STUDY_____ TASK__B-OBTAIN MATERIALS__ DEPARTMENT___CHEMICAL___

	CATEGORY	MONTH 4 - APR			TOTALS TO DATE		
		PLAN	ACTUAL	VARIANCE	PLAN	ACTUAL	VARIANCE
HOURS	SR. PROF.	0	0	0	46	47	⟨1⟩
	JR. PROF.						
	SR. TECH.						
	JR. TECH.						
DOLLARS	LABOR	0	0	0	1,150	1,175	⟨25⟩
	OVERHEAD	0	0	0	1,150	1,175	⟨25⟩
	NONLABOR	0	4,000	⟨4,000⟩	4,400	4,350	50
	PRIME COSTS	0	4,000	⟨4,000⟩	6,700	6,700	0
	G & A	0	600	⟨600⟩	1,005	1,005	0
	TOTAL COSTS	0	4,600	⟨4,600⟩	7,705	7,705	0

Figure 19-7. Cost report for fourth month for task B work by the chemical group.

PROJECT COST REPORT

PROJECT_____MATERIALS STUDY_____   TASK__E-EXPERIMENT_____   DEPARTMENT_TECHNICAL SUPPORT

CATEGORY		MONTH 4 - APR			TOTALS TO DATE		
		PLAN	ACTUAL	VARIANCE	PLAN	ACTUAL	VARIANCE
HOURS	SR. PROF.						
	JR. PROF.						
	SR. TECH.	200	120	80	200	120	80
	JR. TECH.	400	400	0	400	400	0
DOLLARS	LABOR	7,000	5,800	1,200	7,000	5,800	1,200
	OVERHEAD	7,000	7,540	⟨540⟩	7,000	7,540	⟨540⟩
	NONLABOR	0	0	0˙	0	0	0
	PRIME COSTS	14,000	13,340	660	14,000	13,340	660
	G & A	2,100	2,001	99	2,100	2,001	99
	TOTAL COSTS	16,100	15,341	759	16,100	15,341	759

Figure 19-8.   Cost report for fourth month for task E work by the technical support section.

Planned

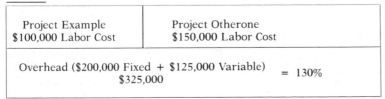

Project Example $100,000 Labor Cost	Project Otherone $300,000 Labor Cost
Overhead ($200,000 Fixed + $200,000 Variable) $400,000 = 100%	

Revised

Project Example $100,000 Labor Cost	Project Otherone $150,000 Labor Cost
Overhead ($200,000 Fixed + $125,000 Variable) $325,000 = 130%	

Figure 19-9.   Cause of unfavorable overhead variance.

direct labor because overhead is partially composed of fixed expenses that cannot be reduced. Thus, the actual overhead rate turns out in this case to be 130 percent rather than the planned 100 percent. (Such a change in overhead is extreme; I use it simply to dramatize the possible effect of overhead rate changes.) The variances could be summarized as follows:

A senior technician was planned full-time but was not released from his previous project at the start of month 4. If labor is not added, the project will be late.

Overhead is now 130 percent, not 100 percent, as planned. This will cause a cost overrun unless compensating savings can be found.

Figure 19-10 is the project cost report for this task in the technical support department at the end of the fifth month. The senior technician category continues to have a favorable variance, which is partially offset by an unfavorable variance in the junior technician category and the continuing unfavorable variance in the overhead. The overhead variance is again attributable solely to the overhead rate now being 130 percent

PROJECT COST REPORT

PROJECT_____MATERIALS STUDY_____ TASK_E-EXPERIMENT_____ DEPARTMENT_TECHNICAL SUPPORT

CATEGORY		MONTH 5 - MAY			TOTALS TO DATE		
		PLAN	ACTUAL	VARIANCE	PLAN	ACTUAL	VARIANCE
HOURS	SR. PROF.						
	JR. PROF.						
	SR. TECH.	40	20	20	240	140	100
	JR. TECH.	80	120	(40)	480	520	(40)
DOLLARS	LABOR	1,400	1,500	(100)	8,400	7,300	1,100
	OVERHEAD	1,400	1,950	(550)	8,400	9,490	(1,090)
	NONLABOR	0	0	0	0	0	0
	PRIME COSTS	2,800	3,450	(650)	16,800	16,790	10
	G & A	420	518	(98)	2,520	2,519	1
	TOTAL COSTS	3,220	3,968	(748)	19,320	19,309	11

Figure 19-10. Cost report for fifth month for task E work by the technical support section.

rather than 100 percent, as per plan. Examination of these variables might lead to the following kind of information:

The senior technician is sick, and this category of labor is below plan.

A junior technician, previously unplanned, has been added and has been able to complete the technician work on schedule.

Overhead is still 130 percent, but final dollars are okay.

The junior technician was able to accomplish in fewer hours the work previously planned for a senior technician (which can indeed happen). This net effective saving in labor hours was sufficient to compensate for the increased overhead dollars. The end result in this case is a small, favorable variance.

TYPICAL PROBLEMS

In addition to management's tendency to presume that cost status is the only measure of a project's progress, a problem previously mentioned, there are two other common problems. First, cost reports are never available immediately after the end of the reporting period (typically the fiscal or calendar month). They are usually issued about two or three weeks after the close of the report period. Thus, there is always a time lag with which the engineering project manager must contend. This problem cannot be solved but must be accepted because of other accounting priorities. An alert project manager can recognize the situation and make full use of other available data to stay more current, including weekly labor distribution reports (indicating who did or did not charge time to the project), purchase orders, travel vouchers, and drawing releases.

The second problem is that cost reports often have errors, such as charges that should have been allocated to other projects or overhead accounts. The project manager or an administrative assistant must study the reports carefully and not merely accept them as gospel.

Cost reports are late and sometimes incorrect.

MATERIALS STUDY PROJECT

> All the examples in this chapter have been based on the materials study project. The ability of computers to sort stored data in a wide variety of ways makes it easy to provide status reports to managers. A senior manager, for instance, Ed Ross, could review other summary information of specific interest to him. Similarly, support team managers could get reports on their subordinates' work.

HIGHLIGHTS *Computer-generated cost reports show variances from plan, which usually require corrective action.*

Reports can show variances due to timing, actual work deviating from plan, or overhead rate changes.

Three problems with cost reports are that management tends to assume they are the only measure of progress, they never appear immediately after the reporting period, and they may contain errors.

FURTHER READING

R. H. Clough and G. A. Sears. *Construction Project Management*, 2nd ed. New York: Wiley-Interscience, 1979.
> *Chapter 10 has an extensive discussion of the project cost system in the construction industry, with much that is generally applicable in any industry.*

J. A. Maciariello. *Program-Management Control Systems*. New York: Wiley-Interscience, 1978.
> *Chapter 11 contains a good review of PERT/COST and C/SCSC cost control approaches.*

R. Pilcher. *Appraisal and Control of Project Costs*. London: McGraw-Hill, 1973.

> *Chapter 11 provides a clear and thorough discussion of cost control, with an emphasis on the need to tie this into the work breakdown structure and network diagram.*

M. D. Rosenau. *Successful Project Management*. Belmont, CA: Lifetime Learning Publications, 1981.

> *Chapter 15, pages 165–179, contains many examples of reports from two computer-based project planning and control systems.*

Handling Changes

Change is a reality of engineering project management. This chapter first reviews why engineering project plans are altered and then discusses techniques for making changes.

REASONS FOR CHANGES

Change is a constant on projects.

Deviations from the project plan occur because Murphy was an optimist. Or there is Rosenau's law of revolting developments: There will be at least one. A large body of data supports this assertion. Table 20-1 summarizes overrun data from a variety of sources. Although not all of these are engineering projects, they all involve technical work. These data convincingly demonstrate that presumably conscientious project managers have been confronted with changes in many situations. This has led to missing both the schedule and cost dimensions of the Triple Constraint, with the overruns typically being a factor of approximately two above plan.

Although not illustrated in Table 20-1, Mansfield's data also show that the situation is somewhat worse than average for large projects and better than average for small projects. This seems to make sense; intuition tells us that the more ambitious undertaking is less likely to be estimated accurately. This is another reason to break a large project down into many small tasks. It will be easier to estimate a small task accurately.

TABLE 20-1. Time and Cost Overrun Data, Expressed As Multiple (X) of Plan.

Project Type	Time	Cost	Source
50 new products (new chemical entities, compounded products, or alternate dosage forms) in ethical drug firm	1.78X	1.61X	E. Mansfield et al., *Research and Innovation in the Modern Corporation*, W. W. Norton, New York, 1971, p. 89
69 new products in proprietary drug laboratory	2.95X	2.11X	E. Mansfield et al., op. cit., pp. 102 & 104
20 management information systems projects	2.10X	1.95X	R. F. Powers & G. W. Dickson, "MisProject Management? Myth, Opinions, and Reality," *California Management Review*, XV, no. 3, 147–156 (Spring 1973)
34 Department of Defense systems from "planning estimate" from "development estimate"	– –	2.11X 1.41X	G. R. McNichols, D.O.D. Report (November 1974), as quoted in R. A. Brown, "Probabilistic Models of Project Management with Design Implications," *IEEE Trans. Engr. Mgmt.*, vol. EM-25, no. 2, 43–49 (May 1978)
10 major construction projects completed 1956–1977	–	3.93X	W. J. Mead et al. (1977), as quoted in E. W. Merrow et al., "A Review of Cost Estimation in New Technologies," Rand Corporation Report R-2481-DOE (July 1979), p. 38
10 energy process plants	–	2.53X	E. W. Merrow et al., op cit., p. 87

Real Changes in the Planned Project

One of the causes of both time and cost overruns is that deviations from the plan occur in the job. They may be externally imposed, by the customer, for instance. They might include a request for increased accuracy in a sensor, a request to add signal indicators on an electronic function generator, or perhaps a request for an additional flight experiment on a space payload.

Changes can be imposed indirectly.

In addition, schedule changes can be imposed, and these frequently have attendant cost implications. If the customer's plant is shut down by a strike or for some other reason the customer does not wish to receive the project output on a stipulated date, there is typically a cost consequence. Conversely, the customer may impose a change in project funding. Although a delay in project funding to the project organization may not appear to change the total available funding, it almost always is accompanied by a schedule rearrangement, which normally leads to undesirable cost consequences.

Environmental, health, and safety regulations that change during the course of an engineering project may cause other changes in scope. Inflation may exceed plan, causing the cost problem, particularly on high-technology aerospace projects originally planned to take several years. There may be changes in resource availability, either people or facilities. These do not constitute changes in project scope, but they do constitute changes from plan that will have an impact, usually unfavorable, on schedule and cost.

Estimation Inaccuracy

Several factors affect estimation accuracy. Of these, an imperfect definition of project scope is the most common. Either a customer or a contractor may be the cause of the error, but it usually is attributable to both.

There may also be poor estimates of either time or cost. The rush to prepare a proposal and submit it in accordance with the bidding requirements may preclude there being sufficient time to do a good job of estimating. There is so much inherent uncertainty in some tasks of some projects that a poor estimate is almost a foregone conclusion.

Uncertainty and "buy-ins" can cause poor estimates.

Many jobs are proposed with deliberate underestimates of the amount of time or money it will take to perform them. This is the so-called "buy-in" situation in which a bidder attempts to win a job by making a low bid. This does not require an illegal misrepresentation, although that may be the case. It may result from a deliberate attempt to make optimistic assumptions about all the uncertainties in the proposed project as well as to omit all contingency from the estimates. In a sense, the bidder is making an estimate of time and cost that could occur perhaps 1 or 0.1 percent of the time rather than attempting to make an estimate near the mid range of possibilities.

Buy-in bids are much more prevalent where the contemplated contract will be a cost reimbursable form and the bidding contractor will not bear the financial burden of having made a low bid. They can also occur in a fixed price contract situation where the bidding contractor is confident that the customer will request changes in scope. Such changes will provide a "get well" opportunity: Increases in both time and cost for the main project can be added onto or concealed in renegotiations necessitated by changes of scope requested by the customer.

ADOPTING CHANGES

At this point, it should be clear that projects normally require changes of plan, although the specific reason or reasons cannot be forecast. The original plan should have included contingency for each dimension of the Triple Constraint, as discussed in Chapter 11. The next chapter reviews several methods to solve the problems caused by the occurrence of revolting developments, stressing the importance of developing and considering alternatives. Once the best alternative has been identified, the change must be adopted.

Figure 20-1 is another variation of a task authorization form, which allows for changes. As I said in Chapter 6, each organization has its own detailed version, but the essential elements are a description of the Triple Constraint, which defines the task being revised, and a place for the person authorizing the revised work and the person accepting responsibility for the revised work to sign. This form thus constitutes a "contract," defining in writing the agreement reached to authorize a revised task. In the case of a subcontractor, a revised or changed subcontract document authorizes the revised task.

Changes, just like originally intended work, must be defined, planned, managed, and monitored before they can be completed. Thus, some, if not all, of the originally issued task

TASK AUTHORIZATION		PAGE
		OF

TITLE			
PROJECT NO.	TASK NO.	REVISION NO.	DATE ISSUED

STATEMENT OF WORK:

APPLICABLE DOCUMENTS:

SCHEDULE

START DATE: COMPLETION DATE:

COST:

	ORIGINATED BY:	DATE:	ACCEPTED BY:	DATE:
ORIGINAL	APPROVED BY:	DATE:	APPROVED BY:	DATE:
	APPROVED BY:	DATE:	APPROVED BY:	DATE:
	ORIGINATED BY:	DATE:	ACCEPTED BY:	DATE:
REVISION 1	APPROVED BY:	DATE:	APPROVED BY:	DATE:
	APPROVED BY:	DATE:	APPROVED BY:	DATE:
	ORIGINATED BY:	DATE:	ACCEPTED BY:	DATE:
REVISION 2	APPROVED BY:	DATE:	APPROVED BY:	DATE:
	APPROVED BY:	DATE:	APPROVED BY:	DATE:

Figure 20-1.   A task authorization form to authorize task revisions.

authorizations must be changed when a change has occurred and a decision has been made as to how to alter the plan to carry out the remainder of the project. This may seem to be a lot of work, but it is far less onerous to take the time to make sure each agreement with people working on the project has been changed than to discover later that some people have been working according to their prior understanding of the project plan.

There is a natural reluctance to make a formal plan change. Such a change not only requires work to issue plan revisions, but it forces us to admit we were wrong (in the original plan), and often brings this to the attention of higher management. Conversely, such information, if given to higher management, may result in your project receiving more help, better access to resources, or higher priority.

Never hesitate to publicize plan changes if these are required.

Figure 20-1 shows a one-page form, but an actual task authorization, in common with a subcontract, might be many pages long. The form might be part of a carbon set, providing copies for the initiator, the task manager, and the project cost accounting section.

Hence, the task authorization documents initially used to authorize work are also a major change control document. A large project may generate many of these, and the amount of time it takes to issue them may be so great as to advise that there first be telephonic or other speedy notice of forthcoming changes.

TYPICAL PROBLEMS

There is always a reluctance to tell the customer and your boss that a revolting development (such as the discovery of unexpected noise in an amplifier) has occurred and many reasons to justify delay. But you should deliver the bad news carefully, thoughtfully, and promptly, before someone else does it.

A second problem is that task authorizations are often verbal rather than written. Because they promote misinterpretation, verbal authorizations should be avoided. But they are employed in the real world of engineering project management. When you must use them, be sure you are clear, ask for feedback, and then try for written confirmation.

The third problem with changes is their impact on resource allocation. There is nothing to do but face up to the reality that resources must be rescheduled, as inconvenient as this may be.

MATERIALS STUDY PROJECT

In the case of the materials study project, NERESCO might request changes if they wanted extra materials to be evaluated, a duplicate experimental setup delivered for their own use, or a formal briefing at the end of the job.

Further changes might arise because a revolting development has occurred. Perhaps the experimental effort (task E) on the materials study project has run into difficulty because one of the materials was found to be impure. If SUPROMAC is anxious to obtain follow-on work from NERESCO, there will be a natural reluctance to inform the customer. Viable alternatives that avoid that necessity would thus be desired. In this instance, properly purified materials can be obtained quickly, and means to recover lost time must then be devised. Two possibilities are using overtime and adding a second shift. If either of these actions can be taken, and the original schedule can be maintained, SUPROMAC need not inform NERESCO.

HIGHLIGHTS   *Changes will occur on every project.*

Changes may result from customer requests; extenuating circumstances, such as a strike; altered environmental, health, or safety regulations; inflation; or resource unavailability.

Uncertainty and buy-ins can result in inaccurate estimations.

Authorization documents can be used to communicate planning and change control.

Three problems changes can cause are managers may be reluctant to inform the customer and higher management of them, verbal authorizations often cause misunderstanding, and resources must be reallocated.

FURTHER READING

R. D. Archibald. *Managing High-Technology Programs and Projects.* New York: Wiley-Interscience, 1976.

Chapter 8, section 2, very briefly discusses change control.

P. W. Metzger. *Managing a Programming Project,* 2nd ed. Englewood Cliffs, NJ: Prentice-Hall, 1981.

Pages 116–120 and 208–209 discuss change control and implementation.

21

Solving the Inevitable Problems

This chapter is concerned with how to solve problems as your engineering project encounters changes during execution. First I discuss some general approaches to coping with problems. Then I describe decision trees, a powerful analytic technique for problem analysis. Following that, I review use of a matrix array and then discuss the kind of meetings in which problem-solving approaches are most likely to be effective.

THE GENERAL APPROACH

Good solutions require a seven-step approach.

In general, the options available are either deductive or inductive logic. In the former, the solution is derived by reasoning from known scientific principles, using analytical techniques, and the conclusions reached are necessary and certain if the premises are correct. In practice, the engineering project manager is rarely confronted with problems for which this approach is appropriate and must rely on inductive techniques, for which the scientific method is the typical prototype. Inductive methods reach conclusions that are probable. This straightforward approach entails seven steps, described in the following sections.

State the Real Problem

The key to problem solving is understanding the real problem rather than the apparent symptoms. Smoke may be emerging from the hardware you built or the computer may refuse to obey a subroutine command, but the actual problem may be an overheated component or an improper line of code in the computer program. You will have to decide how and perhaps why these particular problems occurred.

Gather the Relevant Facts

A fact-gathering phase is usually necessary to clarify the problems. People trained in engineering tend to want to engage in this step ad infinitum. Although it may take a good deal of time to locate information sources, there is also a law of diminishing returns. Because you will never have a 100 percent certainty of obtaining all the information, you must learn to exercise judgment as to when to truncate a search for additional information. At that point, you begin to converge on a solution using the information already gathered.

Propose a Solution

Once a plausible or possible solution has been identified, the winners and losers rapidly separate in their approach to problem solving. The losers inevitably adopt the first solution that comes to mind, possibly leaping out of the frying plan and into the fire. Admittedly, the pressures to come up with a solution quickly are great. No one likes to walk into his or her boss's or customer's office and say, "We have a problem." Such a crisis generates psychological pressure in the project manager to come up with a solution quickly so he or she can say, "We have a problem, but don't worry about it because we have a pretty good solution in mind." But it is best to take a different approach.

Develop Several Alternative Solutions

The winning approach to problem solving is to develop several alternative solutions. Thus, when the problem has arisen and must be reported, the successful project manager will say, "We have a problem. We may have a possible solution, but I am going to take three or four days to consider

Developing alternatives is the key to problem solving.

other alternatives. Then I will report to you on the options and our recommended course of corrective action." Although such an approach to reporting the bad news may make you initially uncomfortable, it is invariably associated with reaching better solutions.

Adopt the Best Alternative

After deciding what is the best alternative, you must adopt a course of action.

Tell Everyone

As an effective project manager, you have earlier made certain that everybody involved in your project knew the original project plan. Now, because you have changed one or more dimensions of the Triple Constraint, you must tell everyone what the new plan is. If you fail to do this, there will be some people working in accordance with obsolete direction, producing something useless and out of date.

Audit the Outcome

As you implement the best alternative solution, watch how it is working out. Auditing will improve your ability to solve problems by showing you how your solutions actually work out. And as you learn more about the problem you are solving and the approach you have adopted, a better alternative may become clear, which may necessitate a further change in the plan.

DECISION TREES

Choosing the best alternative often requires estimating the possible outcomes and their probabilities. An organized way to cope with the situation is to use a decision tree. This technique is both simple and powerful. If you use it often enough, it will improve your average performance in adopting alternatives. Consider the decision whether to go to a movie or walk on the beach. This decision and its possible outcomes are shown in Figure 21-1.

Each possibility has chance future events, which have to do with the quality of the movie or the weather. There are also different outcomes, which are illustrated. Outcomes such as those shown, which have more than one dimension to them (happiness and money, in the figure) must be reduced to a single numeric value. This can be done by utility or preference techniques, which I shall not discuss (see Moore and Thomas, 1976, Chap. 9).

Decision trees require you to quantify the outcome.

Figure 21-2 is a general representation of the kind of decision trees with which you must work. A decision tree always starts with a decision for which there are two or more possible choices. Each choice may be followed by chance future events or subordinate choices (decisions) in any order and with two or more branches following each node. There is a single numeric value outcome, which is typically the present value of the cash flows along the various branches to that outcome.

Decision trees are used routinely in many situations, such as the decision whether to drill an oil well at a given location or whether to locate a new warehouse in a new geographic region. In both of these typical situations, the correctness of the decision depends on the probabilities of things happening in the future (oil being found or business growth), and decision trees are designed to maximize the likelihood of choosing the correct course of action.

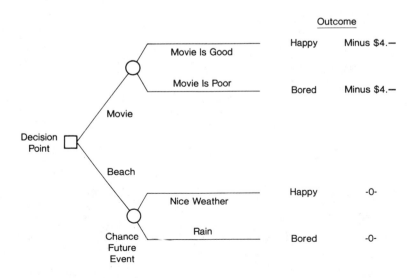

Figure 21-1.   A decision tree.

Consider the following situation, which often confronts an engineering project manager. You have just received an unsolicited request for a proposal that will yield a $300,000 before-tax profit if you can win the job. Checking with the marketing manager, you learn that your company and two others were both solicited suddenly and you all have an equal opportunity of winning the job. Thus, you appear to have one chance in three of winning $300,000 if you write the proposal.

There are two ways to draw decision trees. After discussing this, you and the marketing manager realize you have the opportunity to construct a working model for $45,000 and doing so increases the odds of your winning to fifty-fifty. The decision with which you are confronted is whether to build a working model (at a cost of $45,000) to increase the odds of winning a $300,000 before-tax profit from one in three to one in two. Figure 21-3 shows the decision tree and the problem analysis for this solution. Figure 21-4 is an alternative representation. In both cases, the branch in which the model is built has a higher value than the branch in which the model is not built. Therefore, your decision would be to build the model because it has a higher expected value. Note that this does not guarantee you will win the job. Rather, it

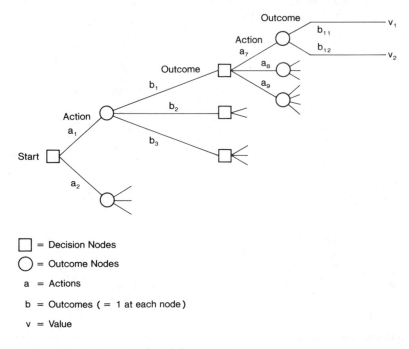

Figure 21-2. A generalized decision tree.

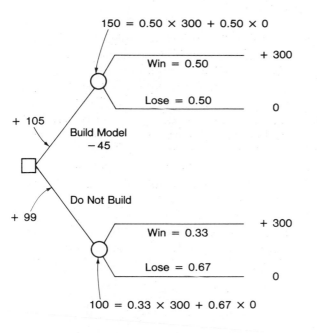

Methodology:
- Construct diagram from left to right.
- Insert $ values from right to left.

Figure 21-3.   Decision tree for illustrative example (thousands of dollars omitted).

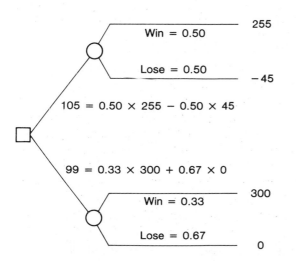

Figure 21-4.   Alternative representation of Figure 21-3 decision tree.

gives you a higher expected dollar value. If you use decision trees in enough cases, you will do better over the long run. But you might very well develop the model and still lose the job, thus ending up losing $45,000. (Note that, for simplicity, I have ignored the cost of writing the proposal in both situations.)

Imagine that you have constructed the preceding decision tree (Figures 21-3 and 21-4), but you and the marketing manager continue to discuss the situation prior to building the model. You realize that, whether or not you build the model, if you lose the proposal, you will have an opportunity to write a subsequent proposal to the winning contractor. Such a subsequent proposal will cost you an additional $10,000, but if you write it after losing, you will have an 80 percent chance to win a subcontract (worth $100,000 in before-tax profit)

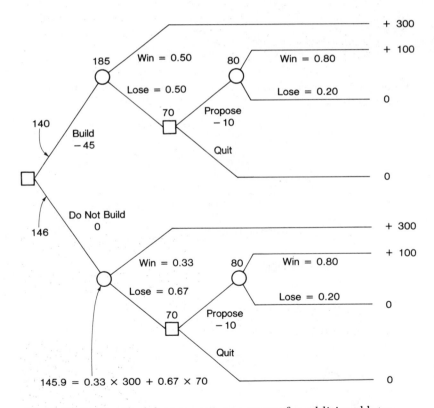

Figure 21-5.   Revised decision tree to account for additional later decision options.

from the winning contractor. In this situation, you would decide not to build the model, as illustrated in Figure 21-5. (Another way to look at this revised situation is to realize that losing the original job is less serious, having an expected value of $70,000 instead of zero.)

Sometimes you are confronted with decision choices, for example, how to remove a resource constraint on two projects (as indicated in Figure 10-4), for which a qualitative decision tree (Figure 21-6) is useful. In this, you may be unable to quantitatively estimate outcome values, odds of occurrence, or costs of actions. However, you may be able to clarify the alternatives to the point where a better choice is possible.

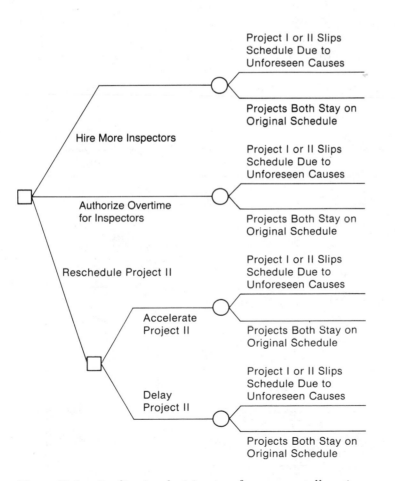

Figure 21-6. Qualitative decision tree for resource allocation.

MATRIX ARRAY

Quantitative

A matrix must list all significant criteria.

In situations where decision trees are not practical or are unwieldy for the analysis and comparison of alternatives, the matrix array may be a satisfactory aid. Figure 21-7 illustrates a quantitative weighting (or scoring) array. The key considerations in the particular problem are listed on the left margin. In a computer programming project, the performance criteria might include processing speed, size of memory required, or similar considerations. For an airborne piece of equipment, performance issues might be weight, size, and reliability. In addition to performance targets criteria, the listing should include the schedule and cost implications of adopting that particular solution.

Next, weighting factors (the sum of which is equal to one) are attached to each criterion (or target). Then a percentage, indicating the degree to which the solution satisfies the particular criterion target, is entered for each solution. Finally, each percentage for each solution in the body of the matrix is

Criteria	Weighting Factor	Solution Approach 1	Solution Approach 2	Solution Approach 3
Performance Target A	0.20	60%	90%	80%
Performance Target B	0.15	90%	70%	70%
Performance Target C	0.15	90%	50%	90%
Schedule Target	0.20	70%	70%	90%
Cost Target	0.30	90%	80%	70%
Weighted Percentage Value		80%	74%	79%

Figure 21-7.   Quantitative decision matrix.

multiplied by the corresponding weighting factors, and the result is entered at the bottom of the solution column (for example, for solution approach 2, 0.20 × 90% + 0.15 × 70% + 0.15 × 50% + 0.20 × 70% + 0.30 × 80% = 74%). There are four problems with this approach:

1. Deciding on weighting factors may be difficult.

2. Choosing percentages may be difficult, especially for the schedule and cost targets, where the solution approach may exceed plan.

3. The highest ranking or weighted percentage value may still be inadequate.

4. It does not consider people, in particular, the possibility that one solution approach is championed by an ambitious person who will work nights and weekends to accomplish it.

Qualitative

Because of the manipulation to which the quantitative approach is subject, it is often better to use a slightly more qualitative matrix. Such a matrix can explicitly consider people issues, especially where there are advocates for or opponents to a particular solution approach. A qualitative matrix can use quantitative data where these are available (see Figure 21-8). Note three aspects of this approach:

1. Numbers are used whenever possible.

2. People are given explicit consideration.

3. The summary identifies both favorable and unfavorable issues.

Also, this kind of matrix can always be used, even early in the decision analysis, before all relevant numerical data are available.

Figure 21-9 shows a qualitative matrix array for the problem of selecting a three-axis "frictionless" support for an airborne instrument to allow five degrees of movement in three directions. The original perceptions of the factors to be considered are typed in the left margin. The first two solutions approaches are listed. Both have many drawbacks. In Figure 21-10, a third alternative solution is identified, pointing out the need to consider one more factor (outgassing). Figure 21-11 shows a reasonably adequate solution for which the schedule must be studied further or perhaps accelerated.

Be honest about solution shortfalls.

To use this kind of matrix, list each solution across the top. You must do this conscientiously. When you write down the first solution and it clearly falls short with regard to one or more of the key criteria, seek additional solutions designed to overcome the shortfall of the ones presently conceived. Usually, after you have written down two, three, or four solutions, each of which has one or more aspects of shortfall, you will identify some hybrid or new variant that comes close to satisfying all the key considerations identified. Even if the matrix array does not lead to a hybrid that clearly satisfies all significant considerations, using the array will clarify available trade-offs and options.

PROBLEM-SOLVING MEETING STYLES

C. J. Margerison, in *Managerial Problem Solving*, identifies four meeting styles for use when solving problems:

1. Command—issue orders
2. Negotiative—different groups horse trading to reach agreement

Issue \ Solution	1	2	3	4	5	6
A						
B						
Schedule						
Cost						
People						
Favorable						
Unfavorable						

Figure 21-8. Qualitative matrix.

Source: "Basic Management Skills for Engineers and Scientists." Lecture notes by Milton D. Rosenau, Jr. Copyright © 1982 by the University of Southern California. Reprinted by permission.

3. Collegiate—peers reaching decision by consensus

4. Advisory—exchange information and make subsequent decision

He also identifies five approaches to problem solving. If the information to solve the problem is readily available, a solution-centered approach to problem solving may be adopted. He provides three of these. If the managerial authority is accepted and the manager has the information and the ability to solve the problem, a directive approach is appropriate and a command meeting, one in which the manager issues orders, is the most appropriate meeting style. This does not mean that a different meeting style cannot be used, but a command meeting will most likely produce an effective solution under these conditions. The negotiative approach to problem solv-

Approach / Consideration	Conventional Outer-Ring Gimbal	Internal Gimbal				
Weight *(Lbs)*	50 ?	20 ?				
Size	Huge	Should Be OK				
Reliability	Bearings	Bearings				
Coercion *(In-Lbs)*	1 ?	1 ?				
Schedule	Quick	Less Quick				
Cost	Low	Higher				
People						
Pro						
Con	Weight, Size, Coercion	Weight, Coercion				

Figure 21-9. Qualitative matrix array.

ing is bargaining with different objectives but common interests, and, in this case, a negotiative meeting is most likely to be effective. The prescriptive approach to problem solving is one in which a solution is solicited, and the presented answer may be only a tentative trial. For this, a negotiative or collegiate meeting style is most useful.

There are two problem-centered approaches to problem solving, those in which information is still needed. The first of these is the consultative approach, in which trust exists and information sharing is useful to diagnose the problem. The second is the reflective approach, which is useful if the problem is unclear and nonjudgmental restatements are acceptable. In both of these approaches, the advisory meeting is most likely to be productive.

Approach / Consideration	Conventional Outer-Ring Gimbal	Internal Gimbal	Gas Bearing			
Weight (Lbs)	50 ?	20?	10-20			
Size	Huge	Should Be OK	OK, But Also Has Pump			
Reliability	Bearings	Bearings	Pump			
Coercion (In-Lbs)	1?	1 ?	Nil			
Outgassing	Oil	Oil	Support Gas			
Schedule	Quick	Less Quick	8-12 Months			
Cost	Low	Higher	Still Higher			
People			Jack			
Pro			Coercion			
Con	Weight, Size, Coercion	Weight, Coercion	Weight, Gas, Reliability, Time			

Figure 21-10. Qualitative matrix array with a third alternative solution identified.

TYPICAL PROBLEMS

In most project management problem-solving situations, it is not possible to find *the* answer, only a most
acceptable (or least objectionable) answer. This may be
caused by the inherent uncertainty or lack of quantitative data. It is thus a matter of judgment about when to
choose among the identified solutions and when to
keep looking for more, better solutions. Honest people
will differ (as they will in their perception of the problem and their evaluation of solution alternatives), and
this must be both expected and tolerated. Use of a
qualitative matrix or a purely qualitative decision tree
can reduce, if not eliminate, this problem.

Approach ⟍ Consideration	Conventional Outer-Ring Gimbal	Internal Gimbal	Gas Bearing	2 Horizontal Flexures + Vertical Mercury	3 Equal (35.2°) Flexures	1 Horizontal + 2 45° Flexures
Weight *(Lbs)*	50?	20?	10-20	∼ 10	5-10	5-10
Size	Huge	Should Be OK	OK, But Also Has Pump	Reasonable	Small	Small
Reliability	Bearings	Bearings	Pump	Mercury Spillage	∞	∞
Coercion *(In-Lbs)*	1?	1?	Nil	∼ 0.01	∼ 0.01	∼0.01
Outgassing	Oil	Oil	Support Gas	Mercury Vapor	Nil	Nil
Schedule	Quick	Less Quick	8-12 Months	3-4 Months	5-6 Months	4-5 Months
Cost	Low	Higher	Still Higher	∼ $15K	∼ $25K	∼ $20K
People			Jack	Ily	Mike	Mike
Pro			Coercion	Time & Cost, Coercion	Weight, Size, Reliability, Coercion	Weight, Size, Reliability, Coercion
Con	Weight, Size, Coercion	Weight, Coercion	Weight, Gas, Reliability, Time	Mercury	Harder to Make Parts	Time = ?

Figure 21-11.   Qualitative matrix array with one reasonably
adequate solution.

MATERIALS STUDY PROJECT

> Considering the materials study project problem from the prior chapter, a qualitative matrix (Figure 21-12) suggests overtime is the least objectionable solution.

Solution / Consideration	Overtime	Second Shift
Performance	OK	OK
Schedule	OK	OK, but must train crew
People	Same as original plan	Need leader and two technicians
Cost	Overtime premium	Shift premium
Pro	People available	Less costly
Con	Overtime premium exceeds shift premium	Extra people training time

Figure 21-12. Qualitative matrix for the materials study project problem cited in Chapter 20.

HIGHLIGHTS   *The seven steps in problem solving are identify the problem, collect the data, devise a solution, search for alternative solutions, adopt the best solution, implement the solution, and audit the outcome.*

Decision trees help you choose the best alternative.

Matrices, another aid to alternative selection, may be qualitative or quantitative.

There are appropriate meeting styles for different problem-solving approaches.

FURTHER READING

V. G. Hajek. *Management of Engineering Projects.* New York: McGraw-Hill, 1977.
> *Section 14.11 is a good one-page summary of decision making to solve problems.*

C. J. Margerison. *Managerial Problem Solving.* Maidenhead, Berkshire, England: McGraw-Hill, 1974.
> *This is an outstanding book on all aspects of problem solving.*

P. G. Moore and H. Thomas. *The Anatomy of Decisions.* New York: Penguin, 1976.
> *This is a well-written, clear, complete treatment of decision trees. It is brief but thorough.*

Part 5

COMPLETING AN ENGINEERING PROJECT

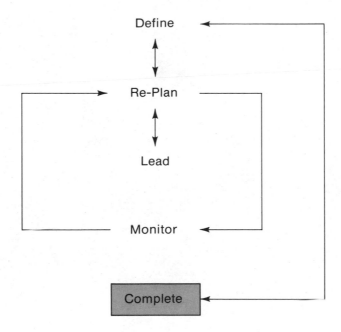

How to Complete an Engineering Project

The fifth and last managerial activity is project completion. To introduce this topic, this chapter first discusses project life cycles. This discussion reviews activities required to complete the engineering project. Then the consequences of project completion are discussed, and it is shown that all personnel do not necessarily have the same stake in ending the project.

PROJECT LIFE CYCLE

Phases

Figure 22-1 shows a project life cycle. The three phases are arbitrary, but they are sufficient to illustrate the point that project activities change during the duration of any engineering project. This has implications for the illustrated personnel head count and means the kind of personnel used in different phases will have to change. For instance, creative designers, very useful in the early phases, can easily become an obstacle to completion if they are retained during the later phases.

That projects have different phases emphasizes that personnel needs will change throughout the project life cycle.

269

Termination

There is a variety of ways to stop projects. Resources can be withdrawn, for instance, by reassigning personnel or required facilities. Higher priority projects may gain at the expense of a low priority project, which may be allowed to wither on the vine. These approaches are not as desirable as an orderly and carefully planned termination. Project success, that is, satisfying the Triple Constraint, can be obtained only by this latter approach. During the last few months of a project, weekly reviews may be required; and during the last few weeks, daily reviews may be required.

Acceptance

The goal of project management is to obtain customer acceptance of the project result. This means the customer agrees that the performance dimension specification of the

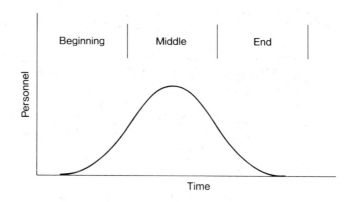

	Beginning	Middle	End
Construction	Site	Erection	Landscaping
Product Development	R&D, Market Research	Engineering and Manufacturing	Product Introduction
Aerospace System	Engineering	Assembly	Customer Test and Sign-off
Computer Software	Definition and Design	Programming	System Test and Acceptance

Figure 22-1.   Project life cycle.

Triple Constraint has been met. Unless the acceptance criteria have been clearly defined in documentation agreed to by the customer and the contractor, there will be discord at the end of the project. When agreement is lacking, the customer will typically want more (unless it costs unacceptable amounts of time or money) and the contractor will argue for less. This is why an unstated or undefined amount of service should not be included in a project.

Never include unspecified service in a project.

Therefore, the acceptance phase must start with the initial contractual definition of the work to be undertaken. This is not to say there cannot be changes of scope during the contract to alter or clarify the acceptance criteria.

In some projects, it may be impossible at the beginning to agree upon final acceptance criteria. This is typical of high-technology development projects. When this is the case, the contract should call for an initial effort of an adequate duration to clarify the entire system design and acceptance criteria. At the end of this first phase, a customer review is conducted and a contract is negotiated for the final phase of work, including acceptance criteria for the end of the project. This approach requires the customer to bear a large risk during the initial phase because the contractor may decide during the initial phase that the final phase will be lengthy and costly. Nevertheless, there is no reasonable alternative to this two-phase approach.

Because there are many possible completion points and delivery conditions for an engineering project, it is necessary to think these through. Completion of the Apollo lunar project would differ depending on whether it was defined as the launch to the moon, the manned landing, the astronaut recovery, or the completion of rock analysis.

Completion requires objective and measurable criteria be attained, which ideally solves the customer's problems.

Objectively measurable criteria for completion (such as test these two amplifiers for gain using a specified method before 9:30 a.m.) are best. Subjective criteria (such as design an attractive housing for this computer) are risky. The former allow little or no room for ambiguity; in the latter, the customer and contractor could easily have different standards.

It is also important to be clear about what the engineering project output is supposed to accomplish. For instance, three very different results occur when the product performs the specified functions, the product was built according to the specified designs, and the product really solved the customer's problem.

Engineering project completion clearly depends upon the precise wording of the acceptance criteria. There should be no room for doubt or ambiguity, although in practice this is

Clear acceptance criteria are required.

extremely difficult to accomplish at project inception. Both parties may enter into a contract with goodwill, but the contractor may have assumed and perhaps even included in the contract the jointly agreed priority to use customer facilities to validate product output. Such a priority can easily become an ambiguous issue toward the end of the project. Other difficulties in wording arise when the word "appropriate" is used. Appropriate tests or demonstrations that seem clear and simple at the beginning of a project have a way of becoming the opposite toward the end.

Delivery

Delivery may or may not be completion. Engineering project completion often requires that the product function after delivery at a location the customer designates. Even if not explicitly the case, there may be an implied warranty that requires it. Thus, responsibility for delivered goods after they leave the contractor's facility is always an issue to be considered at project inception.

Consider the situation where the customer has witnessed final system tests at your company and accepts the system. After delivery to the customer's plant, the system does not perform adequately. What can the project manager do? As a minimum, you are going to have to participate in fact finding. At one extreme, the acceptance criteria will have been absolutely clear; the tests at your company will have been unambiguous; participation in and witnessing of these tests will have been done by qualified and responsible senior personnel; and the performance shortfall at the customer's facility will be attributable entirely to the customer's actions.

More likely, questions will arise about the acceptance test, the specific equipment used to perform it or what might have occurred to change the system during delivery. Regardless of contract form or your desire to obtain more business from that customer, the customer's final happiness with how well the system solves his or her problem will affect your reputation; so you may have to engage in a lot of extra, perhaps unpaid, effort to help get the system working. In some cases, you may be able to show the fault was theirs and get them to pay you for your extra effort.

Once the contractor has turned over possession of the goods, the contractor's control is greatly diminished. Therefore, it is important to have clearly stipulated conditions for acceptance including payment terms.

Documentation Reports

It is not at all uncommon for an engineering project to require the delivery of documentation as well as some other tangible output. Such documentation, in addition to a final report, might include a spare parts list, instruction manuals, and as built drawings. Even the format may be specified.

Because furnishing these items may delay completion of the principal project output, they should be identified as a separate line item in the original contract. Thus, billing for the principal portion of the contract may accompany its delivery and billing for the documentation reports may follow later.

COMPLETION CONSEQUENCES

Project completion may be viewed as a boon or doom. The customer, the contracting organization, the project manager, and the project personnel may not all see it the same way.

Four Affected Parties

For the engineering project manager, completion may be an opportunity for promotion, but many project personnel may find themselves laid off if there is no other work. If the project was badly managed, its manager may receive a less favorable assignment in the future and personnel who did an outstanding job may have a choice future assignment. The contracting organization's view of project completion depends on customer approval and project profitability. The customer may be unhappy because he or she specified the wrong acceptance criteria but admit the contractor has met the specifications.

Everyone does not have the same stake in completion, and the engineering project manager must understand the differences.

Thus, there is no reason to assume that all four parties will have the same view of project completion. Project managers must realize that they may have a very different stake in ending the project than the other three parties. Consequently, the engineering project manager must prod, cajole, or offer inducements to those for whom completion is not obviously desirable.

Completion consequences are also influenced by the reasons for termination. It is certainly best to end the project because all the objectives have been satisfactorily achieved. It is a bad situation if one or more dimensions of the Triple Constraint have been missed substantially.

Personnel Reassignment

Project completion requires reassignment of people. We have now come full circle. The (temporary) project is no longer imposed on the rest of the (permanent) contracting organization. This frequently will necessitate a reorganization of the parent entity because the mix of remaining project work is such that the previously satisfactory organization is no longer appropriate.

A person's perception of what will happen when the project ends will affect his or her work as termination approaches.

The other crucial aspect of personnel reassignment is timing. If a person's next assignment is a choice one, he or she will normally be anxious to start and will lose interest in completing the present project. Conversely, if someone's next project assignment is undesirable, he or she may stall. When no assignment is obvious and layoff or termination is probable, personnel may even attempt sabotage to stretch out the present project assignment.

The engineering project manager can cope with these tendencies to some extent by selecting the time he or she informs project personnel of their next assignment. But if the contracting organization has a reputation for terminating personnel at the end of projects, there is little a project manager can do. The best situation is one in which all project personnel can count on their good work being recognized and appreciated and there being a selection of future assignments.

There are many options for personnel reassignment when projects end.

Even if no specific new project assignment is available when personnel need reassignment, there are still options. For example, personnel can write an unsolicited proposal, prepare an article for publication, work on an in-house development effort, or attend a short course or seminar. Temporary assignments such as these can be used constructively to fill in valleys in the project work load. They also can be used as a motivational tool if they are authorized so as to make participation a mark of recognition for a job well done.

Organizational Changes Due to Completion

When any project ends, for whatever reason, the organization is altered. There is a new body of knowledge in the organization. This is not just tangible information, but also new skills for many people. New working relationships have been established, both within and external to the organization, and these alter the informal organization, even if the formal organization is still unchanged.

INCREASING THE ODDS OF SUCCESS

There are both external and internal factors that influence how well a project satisfies its Triple Constraint.

External Factors

Customers for engineering projects seem to be divided into two broad categories: knowledgeable and shortsighted. The shortsighted customers tend to emphasize the buyer versus seller relationship and to some extent create an adversary relationship between the two organizations. Conversely, knowledgeable customers realize that their stake in project success is ultimately just as great as that of the performing contract organization. Thus, a knowledgeable customer will become involved in the project in an effective, as opposed to a destructive, manner. Such a customer will specify expected reviews and include them in the original job definition. Beyond this, he or she will attempt to ask the tough questions and to carry out probing reviews of the contractor's work, not to embarrass but to assure that all significant issues have been dealt with appropriately. Any required changes will be negotiated intelligently.

A knowledgeable customer, high priority, and clear objectives aid successful completion.

High priority projects inevitably seem to have better outcomes than lower priority projects because they tend to win all competitions for physical and human resources. This is not to say low priority projects lack top management support; top management clearly wants all projects to succeed, but the lower priority projects are at a relative disadvantage.

Clear and stable project objectives are a sine qua non of project success. Objectives can and do change during the course of many projects, but not on a daily or hourly basis. Thus, committing these objectives to writing helps fix them in everyone's mind. Revising them when they must occasionally change is also a requirement of success.

Internal Factors

A qualified, experienced, competent leader is vital, as is a balanced team. Having a team with a balance of skills and getting teamwork from it can be somewhat contradictory. People with very similar backgrounds tend to get along better; so it is easier to promote teamwork in a group composed entirely of, for instance, electrical engineers. Neverthe-

A good leader, a balanced team, the right-sized work packages, careful replanning, and orderly termination contribute to project success.

less, a successful project usually requires a team be composed of more than electrical engineers. Thus, the project manager is confronted with merging people with diverse backgrounds into an effective and harmonious team.

Having the properly sized work packages helps you avoid two potential problems. Complex, difficult work packages should not be assigned to junior people, who may be overwhelmed by them. Simple work packages should not be assigned to senior people, who will not be challenged by them.

Because projects will almost never be carried out exactly in accordance with the original plan, replanning is a constant requirement in project management. Project termination, especially the reassignment of personnel, requires active planning well before scheduled completion.

TYPICAL PROBLEMS

Sometimes not only subordinate personnel but also the project manager must change during the project's life. The manager for the initial phases may be great at the inception but become stale with time or bored by routine wrap-up activities. The solution in this case is to change project managers, and both upper management and the project manager must be alert to this possibility.

MATERIALS STUDY PROJECT

At the end of the materials study project, NERESCO could have asked why more materials had not been tested. In such a case, SUPROMAC must cite the contract and remind NERESCO that an amendment (more materials, more time, more money) could be negotiated to permit this.

Assuming NERESCO is thus satisfied with the number of materials actually tested, SUPROMAC can com-
continued

plete the materials study project by submitting its written report to NERESCO. However, even if that is the only contractual obligation, it may be better to present it as a part of an oral briefing in order to (1) improve communication by allowing some give-and-take discussion and (2) provide an opportunity to increase the customer's happiness with the results.

Although the project life cycle starts with the definition phase and ends with the completion phase, project completion and customer acceptance depend on agreements reached during the definition phase.   **HIGHLIGHTS**

Personnel needs may change throughout the project life cycle.

It is best to end a project because all dimensions of the Triple Constraint have been satisfied.

The project manager must realize that project completion may not be good for all involved parties and plan for an orderly end well in advance of its scheduled time.

Both internal and external factors contribute to project success.

FURTHER READING

V. G. Hajek. *Management of Engineering Projects.* New York: McGraw-Hill, 1977.
> *Chapters 19 and 20 provide a reasonable treatment of the completion phase.*

P. W. Metzger. *Managing a Programming Project,* 2nd ed. Englewood Cliffs, NJ: Prentice-Hall, 1981.
> *Chapters 5, 6, and 7 give a detailed overview of the completion phase for programming projects, covering system test, acceptance, and installation and operation.*

Final Wrap-Up

Projects end with their completion, but there frequently are postcompletion activities that are necessary to the engineering project and that may be viewed as part of it. They are not, however, necessarily part of the basic contract. These are discussed in this chapter.

CONTINUING SERVICE AND SUPPORT

Continuing service and support may be an obligation. If they are, it must be understood who is to pay for them and when. This is often left to be negotiated when the engineering project is completed. Negotiating at this time may be desirable because of an initial inability to see what may be involved, but it leaves a potential Pandora's box at the end of the project.

Some projects require postcompletion service and support.

The contractor should view continuing service and support as an opportunity and not merely as an obligation. His or her employees will be working with the customer's personnel, providing continuing service and support, if it is included in the engineering project. In so doing, they will have informal opportunities to explore ideas with the customer's personnel and hear about real problems the customer is facing. Thus, these contacts provide the basis for future business opportunities.

OWNERSHIP RIGHTS

Patents and special facilities required for contract performance have ownership value. In general, if it is a cost reimbursable contract in which the customer pays all the costs
incurred, ownership rights revert to the customer at the end
of the job. If it is a fixed price contract, ownership rights generally revert to the contractor unless otherwise stipulated in the
initial contract. These can be points of contention unless they
are discussed and clearly resolved in the initial contract. In
any event, patent applications must be filed if any seem jus-
tified. The party of ultimate ownership must expect to pay for
this activity.

Outline patent rights and hiring policies in the contract.

There is also the issue of people. Obviously, neither the
customer nor the contractor owns any people. But "no compete" clauses in recruiting customer personnel into the contractor organization or vice versa are often in the initial
contract. If they were not, the customer might hire the contractor's personnel to perform the continuing service and
support the contracting organization presumed it was going
to perform and be reimbursed for.

AUDITS

There often are postcompletion audits, especially in contracts
performed for governmental entities. For instance, the *Los
Angeles Times* reports (January 13, 1981, pt. 3, p. 16) that a
major aerospace contractor was investigated by a grand jury
because of allegations that it mischarged NASA for work done
between 1977 and 1979.

It is therefore absolutely essential to retain records for the
required duration and even more important to file and document them in an organized and thorough way. Many of the
people who could explain some audited item may no longer
be available when the audit is performed; so the contractor
might lose an important claim if this is not done. Some
portion of the final payment on a contract may be withheld
until the audit is completed. Therefore, a final financial summary may not be possible until the audit is performed, and
this can be months or even years after the other work is
finished.

A project is not complete until the customer pays the bill.

What can you do if your customer does not pay the bill?
The responsibility for obtaining payment usually resides with
the accounts receivable department and is thus outside the

normal purview of the engineering project manager. Nevertheless, an effective engineering project manager may be able to help persuade a slow-paying customer to pay promptly. And the project manager, in considering the possibility of obtaining more work from the customer, will be concerned with the customer's payment history because that can influence the desirability of further work. If further work is desired, the project manager does not want the accounts receivable department to take hostile action against a delinquent account. Thus, the project manager and the accounts receivable department must discuss these issues and mutually decide on the mix of persuasion, cajoling, or legal action to take.

PEOPLE ISSUES

After being assured that all project personnel have been reassigned (or laid off or terminated if necessary), the engineering project manager has two other things to do. First, send personal letters of thanks, appreciation, or praise to project personnel. Second, send a brief wrap-up report to management. It is smart to cite your own successful performance in this.

TYPICAL PROBLEMS

> Basically, the problem here is simply doing it all. The press or excitement of new items to do often leads to the omission of some postcompletion activities. The solution is to recognize your responsibility in getting these things done.

MATERIALS STUDY PROJECT

> In the case of the materials study project, SUPROMAC could use the final report to submit either an oral or a
> *continued*

preliminary written proposal for further work. Assuming it is well received, they could continue to discuss this with NERESCO until a new project is defined and initiated.

Continuing service and support activities may lead to future business opportunities.

HIGHLIGHTS

Contracts often include assignments of patent rights and "no compete" personnel recruitment clauses.

Records must be carefully kept in case there is a postcompletion audit.

Managers should send letters of appreciation to project personnel and a wrap-up report to the boss.

FURTHER READING

R. D. Archibald. *Managing High-Technology Programs and Projects.* New York: Wiley-Interscience, 1976.
  Chapter 10 is a very short recap of some terminal activities.

V. G. Hajek. *Management of Engineering Projects.* New York: McGraw-Hill, 1977.
  Chapter 21 is a very brief recap of a few terminal activities.

M. Silverman. *Project Management—A Short Course for Professionals.* New York: Wiley Professional Development Programs, 1976.
  Chapter 10 briefly discusses phaseout, mostly covering the people issues.

Part 6

OTHER ISSUES IN ENGINEERING PROJECT MANAGEMENT

Other Issues

Small Engineering Projects

Relative to large projects, small engineering projects have both advantages and disadvantages. Being smaller, they are easier to understand and less likely to get into difficulty. But there is less time and money to recover from anything that goes wrong, and they inevitably lack high priority.

SIMPLIFIED MANAGEMENT

Figure 24-1 is a form that can be used to plan and control small engineering projects. It shows the plan for the three research tasks of the materials study project (assuming, for the purpose of illustration, that this is a small task within the research group). The form allows the work breakdown items and the schedule to be entered in either a network diagram or a bar chart form. The overall plan for labor and nonlabor cost may also be entered. As actual costs are incurred, they can be entered and variances noted. In this particular form, the labor categories do not exactly match those previously used, but the form is merely illustrative and could be changed to include any categories. Note also that the form uses months for the time horizon across the top. This could be changed to weeks or any other convenient time frame. An alternate form, which lacks specific cost information, is illustrated in Figure 24-2.

Small projects have a Triple Constraint to plan and control.

PROJECT MANAGER *Tom Richards*
PROJECT NO. *T-84-03* DATE *26 Dec 83*
PRIOR COST *—O—*
START: MO *Jan* YEAR *84* TITLE *Theoretical Work – Materials Study Project*

MONTHS

KEY PROGRAM ELEMENTS	1	2	3	4	5	6	7	8	9	10	11	12
G – Literature Study												
H – Theoretical Study												
J – Final Report												

FINAL REPORT – Scheduled Submission													
Labor Hours	Senior Professional *(Tom Richards)*		120	160	200	160	120						
	Professional *(Jr)*				400	240							
	Technician												
	Total Plan												
	Actual												
Labor Cost, M Dollars *(Incl. OH, Not G&A)*	Plan		6	8	26	17.6	6						
	Actual												
Non-Labor Cost, M Dollars	Plan		1										
	Actual												
Outstanding Commitments, M Dollars													
Total Cost	Plan		7	8	26	17.6	6						
	Actual												
Cumulative Cost, Labor	Plan		6	14	40	57.6	63.6						
	Actual												
Cumulative Cost, Non-Labor	Plan		1	1	1	1	1						
	Actual												
Total Cumulative Cost	Plan		7	15	41	58.6	64.6						
	Actual												

LEGEND
○ TASK
△ DECISION POINT
☐ COMPLETION OR FALLOUT

Page ___*1*___ of ___*1*___

Figure 24-1. Simplified form for small project planning and control.

PROBLEMS

There are four causes for the problems unique to small engineering projects: tight schedules, tight budgets, small teams, and low priority.

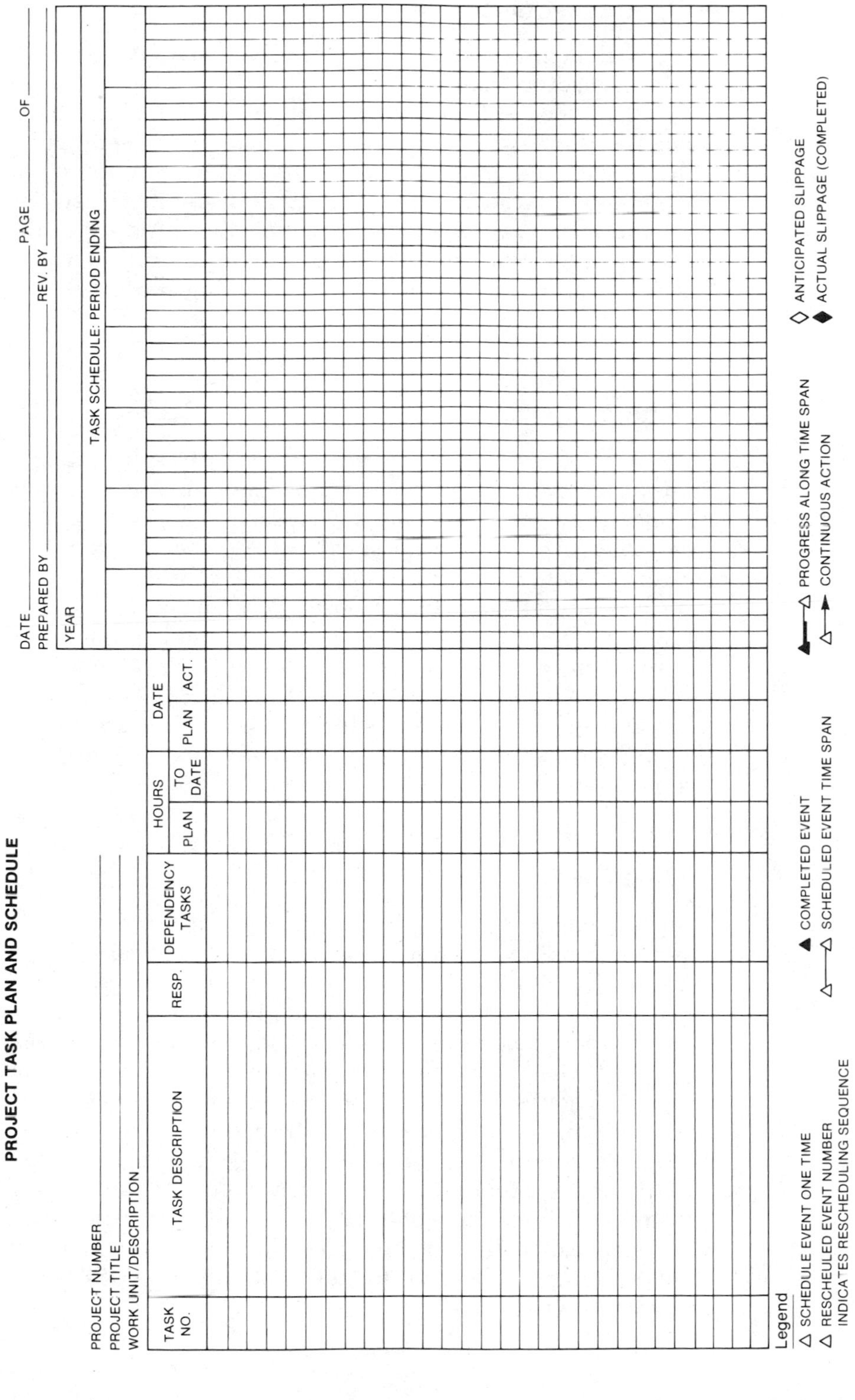

Figure 24-2. An alternate form for small project use.

Tight Schedules

A small engineering project typically is planned to have a shorter schedule than a large project. Thus, the inevitable "getting up to speed" consumes a larger fraction of the available time. A one-week or one-month delay on a long program is less significant than the same delay on a two-, three-, or four-month program, a typical period for a small project. The implication is that small projects must be initiated at the very first opportunity and the project manager must devote a relatively large amount of effort to assuring schedule compliance.

Tight Budgets

Small projects lack time, money, personnel, and priority.

A small engineering project budget will be less than a large project budget. Hence, the absolute amount of money available for contingency must necessarily be less. When the inevitable revolting development occurs, there is less maneuvering room in which to cope with the consequences. The typical two-week lag in cost reports represents a significantly larger fraction of the overall project time and again leads to less reaction capability when some revolting development is discovered. In addition, the cost of time devoted to the program management function is likely to be a larger fraction of the total project budget. Thus, extremely close attention to cost is required on a weekly, if not on a daily, basis.

Small Teams

A large project can typically have the full-time attention of a functional expert (for instance, a thermodynamicist); a small project must make do with the part-time assignment of such specialists. Thus, the small project must compete against other projects for the specialists' time. In some cases, especially where computer programming is called for, this can be a major problem. Each time a person begins a particular programming task, he or she will spend a certain amount of time "getting up to speed." Thus, time is lost in reindoctrination. The problem may be worse with computer programming, but it is not confined to that specialty. Confronted with this reality, the effective engineering project manager should attempt to bargain for full days whenever part-time resources are required.

Priority

Studies have shown that high priority projects are more likely than low priority projects to be completed successfully because they will normally win each competition for any key resource. If you have a low priority engineering project and another person has a high priority project and you both request the model shop make parts for your projects, the person with the high priority project will have his or her needs satisfied first. And it is unlikely that a small project will ever have the same significance for an organization as a large project, which means low priority is more common on small projects than on large projects. Finally, small projects have less visibility, therefore less chance for personal glory; so motivation can be less.

Imagine that you are spending half your time managing a small four-month engineering project and spending the other half working on a much larger project. After two months you discover that the small project is running late and will now require three-quarters of your time to complete it on schedule. There are at least four options to consider:

1. Be late on the small project.
2. Request that you spend only one-quarter time on the large project so you can spend three-quarters of your time on the small one.
3. Request paid overtime approval for the small project.
4. Work unpaid overtime on the small project.

The choice among these, and any other viable options, can be aided by a decision tree or a qualitative matrix.

TYPICAL PROBLEMS

There is another, insidious problem, namely, that of projects starting small and escalating. Again, "staying on top" helps, as does switching to more extensive, formal project management techniques as the project grows.

Beware that small projects do not escalate.

MATERIALS STUDY PROJECT

As shown in Figure 24-1, Tom Richards is to perform the theoretical work on the materials study project (tasks G, H, and J). If he gets ill, quits, or is reassigned to a higher priority project, there is very little time in which to recover (that is, only the slack time until the project's final report, task K).

HIGHLIGHTS   *Four causes of the problems unique to small projects are tight schedule, tight budgets, small teams, and low priority.*

Another problem associated with small projects is a tendency for the project to grow.

FURTHER READING

B. N. Abramson and R. D. Kennedy. *Managing Small Projects.* Redondo Beach, CA: TRW Systems Group, 1975.
> *Although somewhat specific to TRW, this short booklet has some useful pointers.*

M. Zeldman. *Keeping Technical Projects on Target.* New York: AMACON, 1978.
> *This is a graphic manual system that may be helpful for controlling small projects, although the system illustrated in Figure 24-1 seems preferable.*

Where Do You Go from Here?

In this chapter, I wrap up my discussion of the project management process with a summary of the key points you should remember as you put the tools and techniques discussed in this book into practice now or on your next project. Then I mention some trends I feel will impact how you can go about managing projects in the future. Finally, I list some sources for continuing your development of project management skills.

SUMMARY

The following list summarizes the five managerial steps for successful project management:

1. Define
   - Project objectives
   - Statement of work with output specifications
   - Contract

2. Plan, with contingency
   - Performance axis—work breakdown structure
   - Schedule axis—network diagram
     Initial trial
     Activity times estimates
     Critical path calculation
     Adjustment
   - Cost axis—cost estimate by task and performing group
     Each activity on network

3. Lead (or manage)
   • Organization
   • Staffing

4. Monitor (or measure)
   • Reports
   • Reviews
   • Inevitable problems

5. Complete
   • Delivery
   • Acceptance
   • Personnel reassignment
   • Follow-on work

When you start a new project, you should review this list. Then turn to the chapter highlights, where appropriate, for a more detailed refresher. When necessary, you can then review portions of this book or obtain some of the materials cited in "Further Reading."

OTHER SOURCES OF HELP

Sometimes you may still find that you are "over your head." In such a case, you may wish to retain a management consultant for assistance. There are several good sources of qualified consultants. The following associations certify management consulting *companies:*

Association of Management Consultants
Suite 1400
500 North Michigan Avenue
Chicago, IL 60611
   312-266-1261

ACME
230 Park Avenue
New York, NY 10017
   212-687-9693

The following associations certify *individual* management consultants:

Institute of Management Consultants
19 West 44th Street
New York, NY 10036
   212-921-2885

Society of Professional Management Consultants
16 West 56th Street
New York, NY 10019
    212-586-2041

There are other qualified management consultants unaffili-
ated with any of these organizations. However, you have a
better assurance of professional and ethical assistance if you
choose a properly certified management consultant.

THE FUTURE OF PROJECT MANAGEMENT

I think it is likely that more engineering project managers
will begin to have real-time project management status reports (technical, task status, labor hours, costs, and others)
during the next decade. These reports may be provided on
suitable terminals connected to mainframe computers or
they may come from microcomputer networks. Such real-
time status information should allow project managers to
recognize problems more quickly, which can reduce lost time
if the project manager acts promptly.

In addition to real-time information, better graphic output devices (such as laser and color printers) will make it
more convenient to keep network diagrams current. These
diagrams can be distributed electronically and be produced
at remote work sites. Better software will also facilitate
graphical representation of management data that is now
presented numerically, thus highlighting areas of change.

Finally, I think it would be desirable (but less likely) for
organizations to install an adjudicator function. This adjudicator would hear disputes about internal agreements, such as
task work orders, and render a decision. This would institute
inside the organization an analog to the external use of courts
or arbitration to resolve contract disputes. This kind of
mechanism could go a long way to defusing internal corpo-
rate politics and the acrimony that accompanies it.

CONTINUING PROJECT MANAGEMENT SKILL
DEVELOPMENT

Reading this or any other book will not make you an expert
project manager. It takes time and practice for the skills to

become second nature—and you will have to develop your own style, consistent with your skills, interests, and personality. There is no substitute for your own experience. What works for one person may not be useful or appropriate for another. Thus, you should continue to experiment, read, and seek out other sources for continuing education.

Reading

The citations at the end of each chapter can be used to identify other sources available at the time this book was published. More will be published in the future; so you should watch for these.

In addition, you may want to read the *Project Management Quarterly*, published by the Project Management Institute (P.O. Box 43, Drexel Hill, PA 19026; 215-622-1796). This is the only journal devoted to project management, but unfortunately it is mostly dedicated to heavy construction projects.

Continuing Education

A consortium of universities called the Association for Media-based Continuing Education for Engineers (AMCEE, 225 North Avenue, NW, Atlanta, GA; 404-894-3362) has a videocassette course on project management. This course can be leased or purchased.

Alternatively, two sponsors of "live" project management training with which I am familiar are

UCLA Extension (CEEM)
Room 6266
Boelter Hall
University of California
Los Angeles, CA 90024
    213-825-3344 and 825-1295

Industrial Relations Center
California Institute of Technology
383 South Hill Avenue
Pasadena, CA 91125
    213-356-4041

Many other universities and commercial organizations offer seminars of one day to one week duration. These vary in quality, teaching method, and subject matter; so you should determine who will lecture and lead these training programs.

Then try to check out references by talking with prior participants.

The unique value of attending seminars or courses is the
interaction with other project managers. No amount of reading or passive observation of videocassette courses—even if
the materials are outstanding—can give you real practice
with the development of human relationship skills, which are
crucial when you must work with people.

If you cannot attend a university or commercial seminar,
you may want to explore having such a seminar conducted in
your own organization. Although you will not obtain the
stimulus of interacting with personnel from other, different
organizations, you will have a seminar customized to your
own specific situation (perhaps using your own forms as
examples), and usually such an "in-house" seminar can be
much more cost-effective than sending many people to other
seminars.

A FINAL THOUGHT

Good luck! May all your projects be successful.

Appendix 1

Abbreviations Used in Engineering Project Management

ACO	—Administrative Contracting Officer
ACWP	—Actual Cost of Work Performed
AON	—Activity on Node
B & P	—Bid and Proposal
BAC	—Budget at Completion
BCWP	—Budgeted Cost of Work Performed
BCWS	—Budgeted Cost of Work Scheduled
CCB	—Change (or Configuration) Control Board
CCDR	—Contractor Cost Data Reporting
CCN	—Contract Change Notice
CDR	—Critical Design Review
CDRL	—Contract Data Requirements List
CFE	—Customer Furnished Equipment
CFSR	—Contract Funds Status Report
CMO	—Contract Management Office
CPFF	—Cost Plus Fixed Fee
CPIF	—Cost Plus Incentive Fee
CPM	—Critical Path Method
C/SCSC	—Cost/Schedule Control System Criteria
C/SSR	—Cost/Schedule Status Report
CWBS	—Contract Work Breakdown Structure
DAR	—Defense Acquisition Regulations
DDR	—Detailed Design Review

EAC   —Estimate at Completion
ECN   —Engineering Change Notice
ECP   —Engineering Change Proposal
EF   —Earliest Finish
ES   —Earliest Start
ETC   —Estimate to Complete
FCA   —Functional Configuration Audit
FFP   —Firm Fixed Price
FP   —Fixed Price
FQR   —Formal Qualification Review
FQT   —Formal Qualification Testing
G & A —General and Administrative
GFE   —Government Furnished Equipment
IR & D —Internal Research and Development
LF   —Latest Finish
LS   —Latest Start
ODC   —Other Direct Cost
PCA   —Physical Configuration Audit
PDR   —Preliminary Design Review
PERT   —Program Evaluation and Review Technique
PM   —Project (or Program) Manager (or Management)
PMS   —Project Management Systems
PO   —Purchase Order
PR   —Purchase Requisition
PWA   —Project Work Authorization
QA   —Quality Assurance
QC   —Quality Control
RFP   —Request for Proposal
RFQ   —Request for Quotation
SOW   —Statement of Work
T & M —Time and Material
WBS   —Work Breakdown Structure
WO   —Work Order
WU   —Work Unit

Appendix 2

Glossary of Engineering Project Management Terms

Activity A single task within a project

Actual cost of work performed (ACWP) A term in the Cost/Schedule Control System Criteria (C/SCSC) system for the costs actually ·ncurred and recorded in accomplishing the work performed within a given time period

Bar chart A scheduling tool (also called a Gantt chart) in which the time span of each activity is shown as a horizontal line, the ends of which correspond to the start and finish of the activity as indicated by a date line at the bottom of the chart

Bid/no bid decision The decision whether or not to submit a proposal in response to a request for proposal

Bottom up cost estimating The approach to making a cost estimate or plan in which detailed estimates are made for every task shown in the work breakdown structure and summed to provide a total cost estimate or plan for the project

Budgeted cost of work performed (BCWP) A term in the Cost/Schedule Control System Criteria (C/SCSC) system for the sum of the budgets for completed work packages and completed portions of open work packages, plus the appropriate portion of the budgets for level of effort and apportioned effort

Budgeted cost of work scheduled (BCWS) A term in the Cost/Schedule Control System Criteria (C/SCSC) system for the sum of budgets for all work packages, planning packages, and similar items scheduled to be accomplished (including in-process work packages), plus the amount of level of effort and apportioned effort scheduled to be accomplished within a given time period

Burst node In a network diagram, a node at which two or more activities commence after the completion of the preceding activity

Buy-in The process of making a cost bid in a proposal that is unduly optimistic or even actually less than the estimated costs for the project, which is done for the purpose of winning the job

Chart room A room filled with planning documents displayed as charts, typically hung on the walls of the room, used on large projects, and usually marked to indicate current status

Commitment An obligation to pay money at some future time, such as a purchase order or travel authorization, which represents a charge to a project budget even though not yet actually paid

Contingency An amount of design margin, time, or money inserted into the corresponding plan as a safety factor to accommodate unexpected and presently unknown occurrences that judgment suggests will occur during the project (also called reserve)

Costed work breakdown structure A work breakdown structure in which the costs corresponding to major elements of the WBS are also shown on the WBS

Cost plus fixed fee (CPFF) contract A form of contractual arrangement in which the customer agrees to reimburse the contractor's actual costs, regardless of amount, and in addition pay a negotiated fee independent of the amount of the actual costs

Cost plus incentive fee (CPIF) contract A form of contractual arrangement similar to CPFF except that the fee is not preset or fixed but rather depends on some specified result, such as timely delivery

Cost/schedule control system criteria (C/SCSC) A planning and control reporting system devised by the Department of Defense for its contractors to use, intended to foster greater uniformity as well as provide early insight to impending schedule or budget overruns

Critical path In a network diagram, the longest path from start to finish or the path without any slack, thus, the path corresponding to the shortest time in which the project can be completed

Critical path method (CPM) A type of network diagram in which the activities are labeled on the arrows

Customer furnished equipment (CFE) Equipment provided to the contractor doing the project by the customer for the project and typically specified in the contract

Documentation Any kind of written report, including such items as final reports, spare parts lists, instruction manuals, test plans, and similar project information

Dummy activity An activity in a network diagram that requires no work, signifying a precedence condition only

Earliest finish In a network diagram schedule, the earliest time at which an activity can be completed

Earliest start In a network diagram schedule, the earliest time at which an activity can be started

Firm fixed price (FFP) contract   A contractual form in which the price and fee are predetermined and not dependent on cost

Fixed price (FP) contract   Same as firm fixed price

Float   Same as slack time

Functional organization   The form of organization in which all people with a particular kind of skill (such as engineering) are grouped in a common department, reporting to a single manager for that particular functional specialty

Government furnished equipment (GFE)   The same as customer furnished equipment in projects where the customer is a governmental entity

Hardware project   A project in which the principal deliverable item is a product or functioning device of some sort

Latest finish   In a network diagram schedule, the latest time at which an activity can be finished

Latest start   In a network diagram schedule, the latest time at which an activity can be started

Matrix organization   The form of organization in which there is a project management functional specialty as well as other functional specialties and where the project management function has responsibility for accomplishing the project work by drawing upon the other functional specialties as required

Merge node   In a network diagram, a node at which two or more activities precede the start of the subsequent activity

Milestone   A major event in a project, typically one requiring the customer to approve further work

Network diagram   A scheduling tool in which activities or events are displayed as arrows and nodes in which the logical precedence conditions between the activities or events are shown

Periodic review   Any kind of project review conducted on a periodic basis, most commonly a monthly project review

"The Plan"   A document or group of documents that constitutes all the plans for the project, frequently contained in a notebook or series of notebooks

Planning matrix   A matrix in which planned activities are listed on one side (usually the left) and involved people or groups are listed across a perpendicular side (usually the top) and where involvement of a particular individual or group in a particular activity is signified by a tic mark where the row and column intersect

Program   Used interchangeably with "project," as in "program management" or "program manager"

Program Evaluation and Review Technique (PERT)   The form of network diagram in which events are displayed as nodes and connecting arrows indicate the precedence constraints

Progress payments   Payments made to the contractor by the customer during the course of the project, rather than at the end of the project, the terms of which are specified in the contract

Project   An organized undertaking utilizing human and physical resources, done once, to accomplish a specific goal, which is normally defined by a Triple Constraint

Project cost accounting system   A cost accounting system that accumulates actual costs for projects in such a way that total costs for all work in an organization can be allocated to the appropriate projects, normally providing monthly cost summaries; also used in cost planning to summarize the detailed task cost estimates

Project organization   The form of organization in which all or nearly all the people working on a project report to the project manager

Project plan   The entire plan for a project, consisting of the work breakdown structure, network diagram, and task budgets, but sometimes taken to mean only the network diagram

Project team   A term used in this book to designate the personnel working on a project who report to the project manager administratively, not merely for the work on the project

Proposal   A document (sometimes accompanied by models) that an organization submits to a prospective customer that describes work the organization is offering to do

Request for proposal (RFP)   A document issued by one organization to another organization (or to several other organizations) describing work that the issuer wishes to have undertaken by the recipient(s) and inviting the recipient(s) to respond with a proposal

Request for quotation (RFQ)   Similar to a request for proposal, except that the desired items to be procured are stock or catalog items, and only price and delivery time need be proposed

Slack time   In a network diagram, the amount of time on any path other than the critical path that is the difference between the time to a common node on the critical path and the other path

Software project   A project in which the principal deliverable item is a report or other form of documentation, such as a computer program

Statement of work (SOW)   That portion of a proposal or the resulting contract that states exactly what will be delivered and when

Subcontractor   An organization, usually a company, working for another organization on some aspect of the project for which the other organization is under contract

Support team   A term used in this book to designate the personnel working on a project who do not report to the project manager administratively

Task   A small part of a project

Task force An ad hoc group designated to cope with a project, similar to a project organization although frequently staffed with personnel on part-time assignment, usually adopted by a functional organization having only one project or at most a few projects at any given time

Time compression The act of reducing the planned time for an activity, accomplished perhaps by adding unplanned staff or using overtime

Time and material (T&M) contract A contractual form in which the customer agrees to pay the contractor for all time and material used on the project, including a fee as a percentage of all project costs

Top down cost estimating The approach to making a cost estimate or plan in which judgment and experience are used to arrive at an overall amount, usually done by an experienced manager making a subjective comparison of the project to previous, similar projects

Topical review Any kind of project review devoted to a single topic, such as a final design review or a manufacturing review

Triple Constraint The term used in this book to describe the three key project objectives that must be simultaneously accomplished— the performance specification, the time schedule, and the monetary budget

Venture organization The form of organization used in some large organizations where a three- or four-person team, itself functionally organized, is established within the larger organization to develop and commercialize a new product

Work breakdown structure (WBS) A family tree, usually product oriented, that organizes, defines, and graphically displays the hardware, software, services, and other work tasks necessary to accomplish the project objectives

Appendix 3

Planning Checklists for Engineering Project Managers

1. Hardware engineering or study projects
    system review
    system approval
    system test criteria
    system test plan
    detailed hardware specification
    customer furnished equipment
    power requirement plan
    weight control plan
    breadboard design
    breadboard fabrication
    breadboard test
    block diagram
    schematic diagrams
    circuit diagrams
    conceptual design review
    preliminary design review
    critical design review
    final design review
    prototype design
    prototype fabrication
    prototype test
    design freeze
    drawings freeze
    functional designs
    system logic design
    optical design
    mechanical design
    electronic design

thermal design
subsystem hardware implementation
subsystem software implementation
subsystem integration
subsystem review
subsystem approval
subsystem test criteria
subsystem test plan
make/buy decisions
long lead items
special test equipment
commercial test equipment
calibration of test equipment
software tests
data reduction plan
operational software
subsystem cabling
system cabling
installation planning
experimental development plan
support plans
support instrumentation
facilities
training plans
repair facilities and requirements
inspection
preshipment review
customer inspection
customer acceptance
preparation for shipment
shipment
customer support
qualification test
flight acceptance test
launch support
mission support
personnel recruitment
personnel reassignment
documentation
   project plan
   integrated schedule
   functional requirements document
   environmental requirements document
   environmental test specifications
   environmental test procedures
   environmental test reports

interface control
safety plan
configuration control plan
failure mode and effect analysis
reliability and quality assurance plan
development test plan
acceptance test procedure
calibration plan
ground data-handling plan
experiment development plan
expendables consumption
engineering drawings and drawing list
parts list
electronic parts acquisition and screening plan
materials documentation
manufacturing release
periodic reports (for example, monthly)
special reports
final reports
instruction manuals
reviews
manufacturing review
management review
critical design review
preshipment review
internal project reviews
subcontractor progress reviews
customer reviews

2. Programming projects
applications requirements
systems requirements
system inputs
system outputs
detailed architectural design
design specifications
functional specifications
security plan
system test and acceptance specifications
feasibility studies
file and data requirements
cost/benefit analysis
system design
program design
system conversion plan
shipment and delivery

turnover to operations
postimplementation reviews
supplies
training aids
review procedures
hardware requirements
personnel capabilities
milestone reviews
milestone documents
product specifications
project plan
operator instructions
user instructions
library
project index
change control system
data base administration
technical interface manuals
hardware reference manuals
release information
systems reference manual

3. Construction projects
project work order plan
personnel assignment
client review meetings
public involvement meetings
environmental report
meetings with regulatory agencies
population projections
residential load projections
commercial load projections
industrial load projections
field checking
stream gauging records
stream gauging measurements
meteorological records
meteorological measurements
environmental baseline data
environmental sampling programs
ground surveys
photogrammetric surveys
geotechnical explorations
permits
access roads
hydrology

water quality and pollution studies
stability analysis
transient analysis
flood control studies
mathematical modeling
easement and permit drawings and descriptions
wiring diagrams
piping diagrams
construction cost estimates
estimated operation and maintenance costs
life cycle cost analysis
economic studies
value engineering
rate study
notice and instructions to bidders
bid schedules
contract documents
recommendations of awards
shop drawing review
construction surveys
construction engineering
construction observation
construction record drawings

Index